Virtual Networks

Virtual Networks

*Pluralistic Approach for the
Next Generation of Internet*

Edited by
Otto Carlos M.B. Duarte
Guy Pujolle

WILEY

First published 2013 in Great Britain and the United States by ISTE Ltd and John Wiley & Sons, Inc.

ISTE Ltd
27-37 St George's Road
London SW19 4EU
UK

www.iste.co.uk

John Wiley & Sons, Inc.
111 River Street
Hoboken, NJ 07030
USA

www.wiley.com

Library of Congress Control Number: 2013937861

British Library Cataloguing-in-Publication Data
A CIP record for this book is available from the British Library
ISBN: 978-1-84821-406-4

Printed and bound in Great Britain by CPI Group (UK) Ltd., Croydon, Surrey CR0 4YY

Table of Contents

Chapter 2. Virtual Network Interfaces   39
Miguel Elias M. CAMPISTA

**Chapter 3. Performance Improvement and Control of
Virtual Network Elements**   83
Igor M. MORAES

Chapter 8. System Architecture Design 251
Otto Carlos M.B. DUARTE

List of Acronyms

ADAGA   Anomaly Detection for Autonomous manaGement of virtuAl
networks

AMD-V   AMD Virtualization

ARP   Address Resolution Protocol

BGP   Border Gateway Protocol

CSS   Cascading Style Sheets

DMA   Direct Memory Access

dom0   Domain 0

domU   User domain

DoS   Denial of Service

GUI   Graphical User Interface

HTTP   HyperText Transfer Protocol

ICMP   Internet Control Message Protocol

ISP   Internet Service Provider

I/O   Input/Output

IT   Information Technology

IVT   Intel Virtualization Technology

LAN   Local Area Network

LLDP   Link Layer Discovery Protocol

MASR   Memory Allocation, Set and Read

MMU   Memory Management Unit

MUC   Maximum Usage Controller

NIC   Network Interface Card

OS   Operating System

OSPF   Open Shortest Path First

QoS   Quality of Service

RIP   Routing Information Protocol

RTT   Round Trip Time

SLA   Service Level Agreement

SMP   Symmetric Multi Processing

SOAP   Simple Object Access Protocol

SR-IOV   Single Root I/O Virtualization

SSH   Secure Shell

SVG   Scalable Vector Graphics

TCP   Transmission Control Protocol

TLV   Type-Length-Value

ToS   Type of Service

TSC   Time Stamp Counter

UDP   User Datagram Protocol

vCPU   Virtual CPU

VLAN   Virtual Local Area Network

VM   Virtual Machine

VMM   Virtual Machine Monitor

VMS   Virtual Machine Server

VoIP   Voice over IP

VPN   Virtual Private Network

VPS   Virtual Private Server

XML   Extensible Markup Language

ZFG   Zero File Generator

Preface

Currently, there is a strong effort on the part of the research community in rethinking the Internet architecture to cope with its current limitations and support new requirements. Many researchers conclude that there is no one-size-fits-all solution for all user, and network provider, needs, and thus advocate a pluralist network architecture. This new architecture radically changes the Internet because it allows the coexistence of different protocol stacks running at the same time over the same physical substrate. Hence, this book describes a pluralist approach as a new architecture for a post-Internet protocol (IP) environment. This post-IP architecture is mainly based on virtual networking with a piloting system able to cope with the constraints. The piloting system is intelligence oriented and helps to choose the best parameters to optimize the behavior of the network through mechanisms coming from multiagent systems. Indeed, the autonomic-oriented architecture associates with each piece of network equipment (router, box and so on) a situated view that will be used to determine the context and to choose and optimize control algorithms and parameters.

Another very important concept for post-IP networking that we are proposing to use is network virtualization to abstract networks as virtual domains (slices/substrates). A virtual domain represents a coherent functional group of instances of virtual routers rather than physical routers. In this dynamic multistack network, multiple virtual networks coexist on top of a shared substrate. These domains will use the piloting system to distribute physical resources and determine what virtual network will be used by a customer. In this context, a service provider will be able to simultaneously run multiple end-to-end services with different performance and security levels.

Virtual networks can be created and deleted when necessary. Virtualization enables a much better use of physical resources of the network, bringing adapted networks for the customers.

Most of the experimental results presented in this book are from the Horizon Project, a binational research project financed by the Agence Nationale de la Recherche (ANR) in France and Financiadora de Estudos e Projetos (Finep) in Brazil. The consortium is composed of five academic and three industrial partners. The academic partners are UPMC - Paris 6 (Laboratoire d'Informatique de Paris 6 - LIP6), Telecom SudParis, Universidade Federal do Rio de Janeiro (UFRJ), Universidade Estadual de Campinas (Unicamp) and Pontifícia Universidade Católica do Rio de Janeiro (PUC-Rio). The industrial partners are Ginkgo-Networks SA, which has led the work on the knowledge plane, Devoteam, which has worked on convergence infrastructure, and Netcenter Informática LTDA, which has worked on network devices.

Chapter 1, written by Luís Henrique M.K. Costa, discusses virtualization techniques that basically allow us to share computational resources, i.e. to slice a physical computational environment into virtual computational environments that are isolated from each other. This chapter describes the main features and identifies the performance trade-offs of Xen, VMware, and OpenVZ virtualization tools. Performance results of the virtualization tools in terms of resources used by a virtual router – CPU, RAM memory, hard disk and network – are presented. We thank Marcelo Duffles Donato Moreira, Carlo Fragni, Diogo Menezes Ferrazani Mattos and Lyno Henrique Gonçalves Ferraz who have defined the benchmarks and carried out the performance tests.

Chapter 2, written by Miguel Elias M. Campista, presents Xen and OpenFlow virtualization platforms in detail and provides a performance analysis of both. These two platforms were chosen to serve as the basis for new proposals developed in the Horizon project. This chapter also defines the primitives that the network virtualization infrastructure must provide to allow the piloting plane to manage virtual network elements. We thank Natalia Castro Fernandes, Marcelo Duffles Donato Moreira, Lyno Henrique Gonçalves Ferraz, Rodrigo de Souza Couto, and Hugo Eiji Tibana Carvalho who have defined the interfaces and carried out the experiments.

Chapter 3, written by Igor M. Moraes, presents the management tools for the two platforms discussed in the previous chapter. To control and manage virtual network elements, five primitives which the network virtualization infrastructure must provide are defined: instantiate, delete, migrate, monitor and set. One prototype for the Xen platform and another prototype for the OpenFlow platform were designed and developed using the proposed interfaces for both platforms for the sake of proof of concept. We thank Diogo Menezes Ferrazani Mattos, Lyno Henrique Gonçalves Ferraz, Pedro Silveira Pisa, Hugo Eiji Tibana Carvalho, Natalia Castro Fernandes, Daniel José da Silva Neto, Leonardo Pais Cardoso, Victor Pereira da Costa, Victor Torres da Costa, Rodrigo de Souza and Rafael dos Santos Alves who were the main developers of the tools and carried out the experiments.

Chapter 4, written by Edmundo R.M. Madeira and Guy Pujolle, describes context-aware technologies and multiagent systems. The piloting system is based on the multiagent paradigm, developed in a distributed fashion to increase the network scalability. Thus, three platforms for building agents are presented.

Chapter 5, written by Miguel Elias M. Campista, discusses existing control algorithms for virtual networking. This chapter also analyzes the main challenges for packet forwarding using Xen as a virtualization tool and describes in more detail a proposal for local control of virtual networks. Within each physical node, this proposal provides virtual network isolation, guaranteeing the service level acquired by each virtual network, even in the presence of misbehaving virtual networks. The secure virtual network monitor, called XNetMon, described in this chapter was proposed and evaluated by Natalia Castro Fernandes and Otto Carlos Muniz Bandeira Duarte.

Chapter 6, written by Edmundo R.M. Madeira and Nelson Luís S. da Fonseca, describes the piloting system. The idea is to introduce an autonomic system to cope with the increasing complexity of communication networks, releasing the needed network administrators to deal with tasks that require human intervention, such as setting management policies and promoting automation of tasks – system configuration and optimization, disaster recovery and security. A multi-agent self-management prototype is presented. The experiments were carried out by Carlos Roberto Senna and Daniel Macêdo Batista.

Chapter 7, written by Otto Carlos M.B. Duarte, deals with management and control functions. After monitoring and obtaining the usage profile, a knowledge plane uses prediction mechanisms to proactively detect the necessity for updates in the virtual network configuration. The knowledge plane stores information, assists management decisions and executes network maintenance. The fuzzy control scheme was proposed and evaluated by Hugo Eiji Tibana Carvalho and the ADAGA scheme was proposed and evaluated by Pedro Silveira Pisa.

Chapter 8, written by Otto Carlos M.B. Duarte, details the virtualization technologies used for the system architecture. Xen-based routers, OpenFlow switches and a combination of both, called XenFlow, are used to integrate machine and network virtualization techniques. The key idea of XenFlow is to use OpenFlow for managing flows and also for supporting flow migration without packet losses and to use Xen for providing routing and packet forwarding. XenFlow was proposed and evaluated by Diogo Menezes Ferrazani Mattos and Otto Carlos Muniz Bandeira Duarte.

We would like to acknowledge Professors Carlos José Pereira de Lucena, Firmo Freire, Djalmal Zeghlache, Jean-François Perrot, Thi-Mai-Trang Nguyen and Zahia Guessoum. Our thanks also go to Marcelo Macedo Achá and Cláudio Marcelo Torres de Medeiros. We would like to acknowledge the authors of original ideas and papers in Portuguese who are not referenced here but who have introduced concepts discussed in this book. Finally, we also acknowledge all the people who have worked with the Horizon project and have provided many constructive and insightful comments: Alessandra Yoko Portella, André Costa Drummond, Andrés Felipe Murillo Piedrahita, Callebe Trindade Gomes, Camila Patrícia Bazílio Nunes, Carlo Fragni, Carlos Roberto Senna, Claudia Susie C. Rodrigues, Cláudio Siqueira Carvalho, Daniel José da Silva Neto, Daniel Macêdo Batista, Diogo Menezes Ferrazani Mattos, Eduardo Rizzo Soares Mendes de Albuquerque, Elder José Reioli Cirilo, Elysio Mendes Nogueira, Esteban Rodriguez Brljevich, Fabian Nicolaas Christiaan van't Hooft, Filipe Pacheco Bueno Muniz Barretto, Gustavo Bittencourt Figueiredo, Gustavo Prado Alkmim, Hugo Eiji Tibana Carvalho, Igor Drummond Alvarenga, Ilhem Fejjari, Ingrid Oliveira de Nunes, Jessica dos Santos Vieira, João Carlos Espiúca Monteiro, João Vitor Torres, Juliana de Santi, Laura Gomes Panzariello, Leonardo Gardel Valverde, Leonardo Pais Cardoso, Luciano Vargas dos Santos, Lucas Henrique Mauricio, Lyno Henrique Gonçalves Ferraz, Marcelo Duffles

Donato Moreira, Martin Andreoni Lopez, Milton Aparecido Soares Filho, Natalia Castro Fernandes, Neumar Costa Malheiros, Nilson Carvalho Silva Junior, Othmen Braham, Pedro Cariello Botelho, Pedro Henrique Valverde Guimarães, Pedro Silveira Pisa, Rafael de Oliveira Faria, Rafael dos Santos Alves, Raphael Rocha dos Santos, Renan Araujo Lage, Renato Teixeira Resende da Silva, Ricardo Batista Freitas, Rodrigo de Souza Couto, Sávio Rodrigues Antunes dos Santos Rosa, Sylvain Ductor, Tiago Noronha Ferreira, Tiago Salviano Calmon, Thiago Valentin de Oliveira, Victor Pereira da Costa and Victor Torres da Costa.

Otto Carlos DUARTE
and Guy PUJOLLE
June 2013

Chapter 1

Virtualization

In this book, we focus on a novel Internet architecture based on the pluralistic approach. An example of a pluralistic architecture is shown in Figure 1.1. In Figure 1.1, each router layer represents a different network with independent protocol stacks that share the resources from the underlying network infrastructure at the bottom layer. Virtualization is a key technique to make such a pluralistic architecture possible. Virtualization is a technique that basically allows sharing of computational resources [POP 74]. Virtualization divides a real computational environment into virtual computational environments that are isolated from each other and interact with the upper computational layer as would be expected from the real non-virtualized environment. A comparison between a virtualized and a non-virtualized environment is shown in Figure 1.2. The left-hand side of the figure shows a traditional computational environment where applications are executed on top of an Operating System (OS) that controls the underlying hardware. On the right-hand side of the figure, a virtualized environment is shown, where a virtualization layer allows multiple OSs to run concurrently, each with its own applications, and control their access to the hardware. As we deal with virtual *networks*, we consider router resources, such as the processor, memory, hard disk, queues and bandwidth, as the computational environment to be virtualized. A set of virtual routers and links is called a *virtual network*.

Chapter written by Luís Henrique M.K. COSTA.

Therefore, using the virtualization technique, we can have multiple concurrent virtual networks, each with a particular network protocol stack, sharing a single physical network infrastructure, as shown in Figure 1.1.

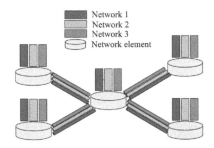

Figure 1.1. *Pluralistic architecture example*

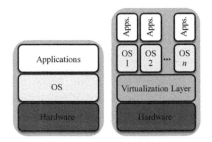

Figure 1.2. *Virtualized environment example*

Virtualization is commonly implemented by a software layer called a hypervisor, which is responsible for multiplexing the computational resources between the multiple virtual environments or Virtual Machines (VMs). Each VM runs over the hypervisor, which controls the access to the physical resources. Different hypervisors and virtualization techniques exist. In this chapter, we give an overview of the main features and identify the performance trade-offs of the most prevalent virtualization tools: Xen [BAR 03, CHI 08], VMware [VMW 07a], and OpenVZ [KOL 06]. The study compares virtualization tools regarding their performance with respect to a virtual router's resources of interest: central processing unit (CPU), random access memory (RAM) memory, hard disk and network. CPU is used by the virtual routers to process incoming packets and route them based on the forwarding tables. RAM is used to store the forwarding tables. The main use of hard disk is to store the VM images. Network resources are used to forward packets, the main task of a router. For normal operation of virtual routers, CPU, RAM and network are the most sensitive resources to virtualization overhead. Disk performance overhead is of interest because it impacts the instantiation of new routers and migration of virtual routers. To better understand the overhead introduced by such tools, native performance is also presented whenever applicable.

Since virtualization can cause malfunctions in time-sensitive applications [VMW 08], we used a different technique than the technique used in related work [XEN 07, VMW 07b]. We based our results on the time taken for a virtualized system to accomplish a task measured from an external non-virtualized computer.

We have conducted two types of experiments. The first type of experiment aim at analyzing the performance loss incurred by the extra layer – the hypervisor – introduced by virtualization tools. To achieve this goal, in the first type of experiment there is only one VM running over the virtualization software. We compare the performance of native Linux with Xen, VMware and OpenVZ virtualization software. Our results show that Xen is a good fit for PC-based router virtualization, having acceptable virtualization overhead, as demonstrated in section 1.4.6, allowing hypervisor modification, for being open source, and providing virtual router flexibility, since it has a virtual hardware interface that allows the use of different OSs in different virtual routers. In addition, single VM experiments provide a baseline for the second type of experiment, which deals with multiple VMs. The second set of experiments investigates how the chosen virtualization tool scales with the number of VMs running in parallel. These kind of tests have the objective of clarifying how instantiating multiple VMs, consuming the same resources, affect the overall performance and how fairness is handled by the virtualization tool. In this kind of test, we also verify how different schemes of CPU cores' allocation among VMs influence the overall performance.

In section 1.1, we present the techniques used by the virtualization tools we tested. In section 1.2, we describe in more detail the virtualization tools. Section 1.3 describes the methodology of the tests and the presents our testbed. Section 1.4 presents the benchmarks used in the virtualization tools comparison, and in sections 1.4.6 and 1.4.7, we present the performance comparison results. Finally, concluding remarks of this chapter are presented in section 1.5.

1.1. Virtualization techniques

To better understand the compared virtualization tools, it is important to identify the different virtualization techniques. This section describes the concepts and techniques used by Xen, VMware and OpenVZ.

There are many different definitions of virtualization but they all agree that virtualization is a technique for sharing computational resources, granting some level of isolation between the virtual environments. The classical definition, according to Popek and Goldberg [POP 74], is that virtualization is a technique that provides VMs, which are efficient isolated copies of the underlying hardware. Today, this concept can be generalized not only on hardware but on any computational resource, as an OS kernel or the VM abstraction used by programming languages such as Java and C#.

Several challenges arise from the virtualization objectives. The first challenge is scheduling all the virtual environments to access the same underlying computational resources. For hardware virtualization, the shared resources are CPU, RAM, storage and network. RAM-shared access can be done in different ways. The virtual environment can be given access to a virtual memory space that can be translated into physical RAM, the same way OSs do. Another approach is to let the VMs be aware of their virtualized nature and allow them to directly access the memory after being assigned an area memory slice by the hypervisor. CPU sharing can be made in several ways and can be achieved using mechanisms such as round-robin, weighted round-robin, allocation under demand and others. Input/output (I/O), in general, can be handled in a unified fashion using buffers for storing exchanged data that is multiplexed and demultiplexed between the physical and virtual peripherals.

Our experiments use commodity x86-based hardware, which imposes additional challenges to virtualization. In the beginning of virtualization development, in the 1970s, hardware was designed to be virtualization enabled. Mainframes had instruction sets in which all the instructions that handled resource allocation and usage were privileged, that is the program required a certain CPU privilege execution level. In these CPU architectures, hardware virtualization could be achieved using a technique called "de-privileging" [ADA 06], where the virtualized OS is executed in an unprivileged context in order to generate traps whenever resources' allocation or usage instructions would be executed. For this reason, the hypervisor would intercept the traps and emulate the allocation or usage of the required resource in a way that is safe for the other VMs. According to Popek and Goldberg [POP 74], there are three requirements for building a hypervisor for hardware virtualization: (1) efficiency, which means that a large subgroup of the instruction set from the virtual CPU (vCPU) should be directly executed in the real CPU without any kind of intervention from the hypervisor; (2) resource control, which means that the hypervisor must have complete control of all the resources; and (3) equivalency, which means that the hypervisor must provide a virtual interface to the virtual environment equivalent to the original interface. The 1970s mainframes would facilitate building hypervisors, since VM isolation could be reached using the de-privileging technique. For x86-based hardware, this is not true, since for optimization purposes x86-based hardware instruction sets have instructions that access shared resources but do not require a privileged context and, moreover, the x86-based instruction sets contain a group of instructions classified as sensitive to the privilege level, in which the instruction is executed in a different manner depending on the current privilege level. If a de-privileged OS executed a sensitive instruction, it would fail silently, since it would not generate a trap nor would it execute in the way which the OS intended. To handle these x86-based hardware problems, there are several workarounds that will be presented in the following sections that describe the different virtualization techniques.

1.1.1. *Full virtualization*

Full virtualization is a virtualization technique in which all the original interfaces are virtualized and the interfaces exposed to the virtual environment are identical to the original interface. In this approach, the guest OS, that is the OS residing inside the VM, does not need to be modified and directly executes inside the VM. To deal with sensitive instructions from the x86-based hardware platforms, different techniques can be used. A well-known technique is binary translation. Binary translation checks the code to be executed, searching for problematic instructions and replacing them with instructions that emulate the desired behavior. The advantage of binary translation is that it allows applications and OSs to be used without modifications. Nevertheless, binary translation incurs high CPU overhead since all executed code must be checked and problematic instructions are replaced in run-time.

Recently, there has been a strong effort from major hardware manufacturers to optimize virtualization. Server consolidation uses virtualization to replace several servers with idle capacity into a single hardware with higher utilization where several virtual servers execute. Server consolidation has become common practice to cut down equipment and maintenance costs in major companies. For this reason, both AMD and Intel have developed technologies for more efficient virtualization support in modern CPUs. Intel Virtualization Technology (IVT) and AMD Virtualization (AMD-V) have both provided better performance for full virtualization by introducing two new operation modes: root and non-root. The root operation mode is used by the hypervisor and is similar to regular CPU operation, providing full CPU control and the traditional four rings of privilege levels. The non-root mode is meant for the execution of the VMs. In this mode, the CPU also provides four rings of privilege levels and the guest OS no longer executes in a de-privileged ring but in ring 0, for which it was designed. Whenever the guest OS executes a problematic instruction, the CPU generates a trap and returns control to the hypervisor to deal with this instruction. With that CPU support, binary translation is no longer necessary and full virtualization hypervisors have greatly increased their performance.

1.1.2. *Paravirtualization*

Paravirtualization is a virtualization technique in which the guest OS cooperates with the hypervisor to obtain better performance. In paravirtualization, the guest OS is modified to call the hypervisor whenever a problematic instruction is executed. The sensitive instruction is replaced by a virtualization-aware instruction that calls the hypervisor. For this reason, the hypervisor does not need to monitor the VM execution for problematic instructions, which greatly reduces overhead in comparison to full virtualization using binary translation. The trade-off is that OS must be modified and recompiled, to produce paravirtualized OS images. The paravirtualized OS image is needed to make hypervisor calls. This hinders virtualization of legacy OSs and requires the cooperation of the OS developer.

1.2. Virtualization tools

In this section, we describe the main virtualization techniques and we present performance evaluation results for Xen, VMware and OpenVZ.

1.2.1. *Xen*

Xen is an open-source hypervisor proposed to run on commodity hardware platforms that uses the paravirtualization technique [BAR 03]. Xen allows us to simultaneously run multiple VMs on a single physical machine. Xen architecture is composed of one hypervisor located above the physical hardware and several VMs over the hypervisor, as shown in Figure 1.3. Each VM can have its own OS and applications. The hypervisor controls the access to the hardware and also manages the available resources shared by VMs. In addition, device drivers are kept in an isolated VM, called Domain 0 (dom0), in order to provide reliable and efficient hardware support [EGI 07]. Because dom0 has total access to the hardware of the physical machine, it has special privileges compared with other VMs, referred to as User domains (domUs). On the other hand, domUs have virtual drivers, called front-end drivers, which communicate with the back-end drivers located in dom0 to access the physical hardware. Next, we briefly explain how Xen virtualizes each machine resource of interest to a virtual router: processor, memory and I/O devices.

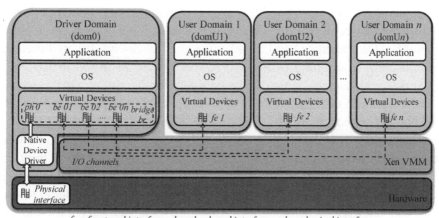

fe – front-end interface   be – back-end interface   ph – physical interface

Figure 1.3. *The Xen architecture*

Xen virtualizes the processor by assigning vCPUs to VMs. vCPUs are the CPUs that the running processes within each VM can see. The hypervisor maps vCPUs to physical CPUs. Xen hypervisor implements a CPU scheduler that dynamically maps a physical CPU on to each vCPU during a certain period. The default scheduler of the used version (3.2) of Xen is the credit scheduler, which makes a proportional CPU share. The credit scheduler allocates CPU resources to each VM (or, more specifically, to each vCPU) according to the weights assigned to VMs. The credit scheduler can also be work-conserving on Symmetric Multi Processing (SMP) hosts. This means that the scheduler permits the physical CPUs to run at 100% if any VM has work to do. In a work-conserving scheduler, there is no limit on the amount of CPU resources that a VM can use.

Memory allocation in Xen is currently done statically. Each VM receives a fixed amount of memory space that is specified at the time of its creation. In addition, to require a minimal involvement from the hypervisor, VMs are responsible for allocating and managing the corresponding portion of the hardware page tables. Each time a VM requires a new page table, it allocates and initializes a page from its own memory space and registers it with the Xen hypervisor, which is responsible for ensuring isolation.

In Xen, data from I/O devices are transferred to and from each VM using shared-memory asynchronous buffer descriptor rings. The task of Xen hypervisor is to perform validation checks, for example checking that buffers are contained within a VM memory space. Dom0 access I/O devices directly by using its native device drivers and also performs I/O operations on behalf of domUs. On the other hand, domUs use their back-end drivers to request device access from dom0 [MEN 06]. A special case of I/O virtualization is network virtualization, which is responsible for demultiplexing incoming packets from physical interfaces to VMs and also for multiplexing outgoing packets generated by VMs. Figure 1.4 illustrates the default network architecture used by Xen. For each domU, Xen creates the virtual network interfaces required by this domU. These interfaces are called front-end interfaces and are used by domUs for all of its network communications. Furthermore, back-end interfaces are created in dom0, corresponding to each front-end interface in a domU. To exchange data between back-end and front-end interfaces, Xen uses an I/O channel, which uses a zero-copy mechanism. This mechanism remaps the physical page containing the data into the target domain [MEN 06]. Back-end interfaces act as the proxy for the virtual interfaces in dom0. Front-end and back-end interfaces are connected to each other through the I/O channel. In Figure 1.4, back-end interfaces in dom0 are connected to the physical interfaces and also to each other through a virtual network bridge. This default architecture used by Xen is called bridged mode. Thus, both the I/O channel and the network bridge establish a communication path between the virtual interfaces created in domUs and the physical interface.

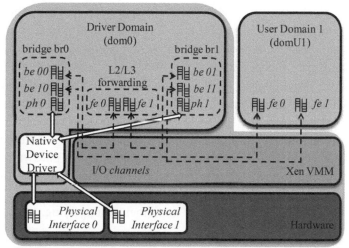

fe – front-end interface , be – back-end interface , ph – physical interface

Figure 1.4. *Xen network architecture*

1.2.2. *VMware*

VMware is a company that provides machine virtualization platforms for end user and datacenter customers. VMware virtualization platforms are based on the full virtualization concept. This work evaluates a VMware datacenter class virtualization platform called VMware ESX Server. It is mainly used for server consolidation and it is one of the most used piece of enterprise virtualization software. VMware ESX Server aims at guaranteeing VM isolation and resource sharing fairness based on resource allocation policies set by the system administrator. Resource sharing is dynamic, because resources can be allocated and reallocated to VMs on demand [VMW 05]. Although this work evaluates the VMware ESX

Server 3.5, the pieces of information reported here are from VMware ESX Server 2.5 due to the fact that VMware ESX Server is a proprietary product and there is little information about its implementation. The description below takes into account version 2.5, considering that there are no significant changes between versions 2.5 and 3.5.

VMware architecture, as shown in Figure 1.5, is composed of the hardware interface components, the virtual machine monitor (VMM), the VMkernel, the Resource Manager and the service console. The hardware interface components are responsible for implementing hardware-specific functions and create a hardware abstraction that is provided to VMs. It makes VMs' hardware independent. The VMM is responsible for CPU virtualization, providing a vCPU to each VM. The VMkernel controls and manages the hardware substrate. VMM and VMkernel together implement the virtualization layer. The Resource Manager is implemented by VMkernel. It partitions the underlying physical resources among the VM, allocating the resources for each VM. VMkernel also implements the hardware interface components. The service console implements a variety of services such as bootstrapping, initiating execution of virtualization layer and resource manager and runs applications that implements supporting, managing and administrative functions.

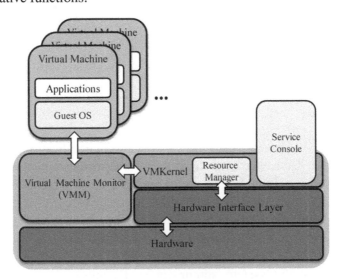

Figure 1.5. *VMware architecture*

VMware ESX Server, as other virtualization tools, virtualizes four main resources: CPU, memory, disk and network device. We will further detail below how VMware virtualizes each resource.

CPU virtualization is done by setting a vCPU for each VM. The VM does not realize that it is running over a vCPU, because vCPUs seem to have their own registers and control structures [VMW 05]. A VM can have one or two vCPUs. When it has more than one CPU, it is called an SMP VM. The VMM is responsible for CPU virtualization by setting system states and executing instructions issued by the VM.

In a virtualized environment, the guest OS runs in a lower privileged level than it was designed to. As mentioned in section 1.1, a classical approach to virtualize CPU resources is trap-and-emulate, which is a technique in which the VM tries executing an instruction, and, if it cannot be executed in a lower privilege level, the CPU generates a trap that is treated by the VMM, which emulates the instruction execution to the guest OS. Nevertheless, as mentioned in section 1.1, this technique does not work for the x86 architecture, since it has instructions that are sensitive to privilege level and would execute in a different way than the OS meant, without generating a trap. To solve this issue and keep a satisfactory performance, VMware combines two CPU virtualization techniques: direct execution and CPU emulation. The instructions from the user-space of a VM are executed directly on the physical CPU, a technique known as direct execution. Instructions that are sensitive to privilege level are trapped by the VMM, which emulates the instruction execution, adding performance overhead. Combining both techniques allows CPU intensive user-space applications to have near-native performance. The performance loss depends on the number of sensitive instructions that had to be replaced.

CPU scheduling is made by the Resource Manager. CPU scheduling is based on shares, which are units used to measure how much time is given to each VM. CPU scheduling is proportional-share, meaning that CPU time given to each VM is proportional to the amount of shares it has in comparison with the total amount of shares in the system. In an SMP VM, CPU allocation is different. Resource Manager schedules the vCPUs one-to-one onto physical CPUs, and tries executing them at the same time. VMware CPU scheduling tries to keep fairness between VMs' CPU allocation. When a VM is halted, or idle, Resource Manager schedules its CPU time to other VMs that are running.

VMware's memory virtualization approach is to create a new level of memory address translation. It is done by providing each guest OS a virtual page table that is not visible to the Memory Management Unit (MMU) [BAR 03]. Within a VMware virtualization environment, the guest OS accesses a virtual memory space provided to the VM. The guest OS page table maintains the consistency between guest virtual pages and guest virtual "physical" pages. Guest virtual pages are virtual memory pages within a VM, as in a native OS virtual memory. However, a guest virtual paging mechanism cannot access directly the physical memory, it accesses guest virtual "physical" memory. Guest virtual "physical" memory is an abstraction of the physical memory. When a guest OS tries to execute an instruction to access physical memory, this instruction is trapped by the VMM and its address is translated to the real physical address. Guest virtual "physical" memory is always contiguous, but can be mapped onto non-contiguous real physical memory. VMware memory sharing obeys administration policies, which, for example, define a minimum and a maximum amount of physical memory to be accessed by a VM. It is also possible to have VMs consuming more than the total amount of physical memory available on the physical machine. This is possible because the host system can also do a swap, as in a traditional virtual memory mechanism used in modern OSs. The memory sharing scheduler works like the CPU scheduler, but takes into account memory shares instead of CPU shares.

VMware's I/O virtualization approach is to emulate performance-critical devices, such as disk and network interface cards. Device accesses are emulated by the VMkernel. The VMkernel calls the hardware interface layer, which is responsible for accessing the device driver and executing the operation on the physical hardware device. For storage virtualization, a Small Computer System Interface (SCSI) driver is presented to the VM. VMs access this driver, and the VMkernel traps driver access instructions and implements VM disks as files in the host file system.

Concerning network I/O virtualization, VMware implements the *vmxnet* [VMW 05] device driver, which is an abstraction of the underlying physical device. When an application wants to send data through the network, the guest OS processes the request and calls the vmxnet device driver. The I/O request is intercepted by the VMM and control is transferred to VMkernel. VMkernel is independent of the physical device. It processes the request, manages the various VM requests and calls the hardware interface layer, which implements the specific device driver. When data arrive to the physical interface, the mechanism for sending it to the specific VM is the same, but in reverse order. The main overhead introduced by this mechanism in the context switching between the VM and the VMkernel. To decrease the overhead caused by context switching, VMware ESX Server collects clusters of sending or receiving network packets before doing a context transition. This mechanism is only used when packet rate is high enough to avoid increasing packet delays.

To conclude, VMware ESX Server is a full virtualization tool that provides a number of management and administrative tools. VMware ESX Server is focused on datacenter virtualization. It provides a flexible and high-performance CPU and memory virtualization. However, I/O virtualization is still an issue, since it is done by emulating the physical devices and involves context changes.

1.2.3. OpenVZ

OpenVZ is an open-source OS-level virtualization tool. OpenVZ allows multiple isolated execution environments over a single OS kernel. Each isolated execution environment is called a Virtual Private Server (VPS). A VPS looks like a physical server, having its own processes, users, files, Internet Protocol (IP) addresses, system configuration and providing full root shell access. OpenVZ claims to be the virtualization tool that introduces less overhead, because each VPS shares the same OS kernel, providing a high-level virtualization abstraction. The main usages for this virtualization technology are in web hosting, providing every customer with a complete Linux environment, and in information technology (IT) education institutions, providing every student with a Linux server that can be monitored and managed remotely [SWS 05]. Despite the small overhead introduced by OpenVZ, it is less flexible than other virtualization tools, such as VMware or Xen, because OpenVZ execution environments have to be a Linux distribution, based on the same OS kernel of the physical server.

OpenVZ architecture, as shown in Figure 1.6, is composed of a modified Linux kernel that runs above the hardware. The OpenVZ-modified kernel implements virtualization and isolation of several subsystems, resource management and checkpoints [KOL 06]. In addition, I/O virtualization mechanisms are provided by the OpenVZ-modified kernel, which has a device driver for each I/O device. This modified kernel also implements a two-level process scheduler that is responsible for, on the first level, defining which VPS will run and, on the second level, on deciding which VPS process will run. The two-level scheduler and some features that provide isolation between VPSs form the OpenVZ virtualization layer. VPSs run above the OpenVZ virtualization layer. Each VPS has its own set of applications and packages, which are segmentations of certain Linux distributions that contain applications or services. Therefore, a VPS can have its own services and applications independent of each other.

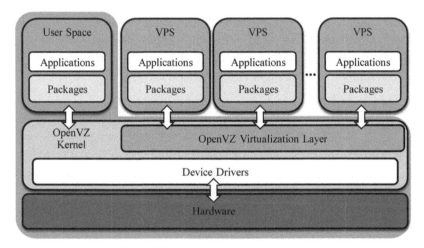

Figure 1.6. *The OpenVZ architecture*

Resource virtualization in OpenVZ is done by allowing or prohibiting a VPS to access a resource on the physical server. In general, resources in OpenVZ are not emulated, they are shared among VPSs. To define the amount of each resource that is guaranteed for each VPS, a number of counters (about 20) are defined in the VPS configuration file. Next, we further detail how OpenVZ virtualizes the processor, memory, disk and network devices.

For processor virtualization, OpenVZ implements a two-level CPU scheduler [KOL 06]. On the first level, the virtualization layer decides which VPS will execute for each time slice, taking into account the VPS CPU priority, measured in *cpuunits*. On the second level, which runs inside the VPS, the standard Linux scheduler defines which process will execute for each time slice, taking into account the standard process priority parameters.

OpenVZ allows VPSs to directly access the memory. Moreover, it is more flexible than other virtualization technologies, such as Xen. During VPS execution, the memory amount dedicated to one VPS can be dynamically changed by the host administrator. OpenVZ kernel manages VPSs memory space to keep in physical memory a block of the virtual memory corresponding to the running VPS.

OpenVZ virtual disk is a partition of the host file system. Similarly to CPU scheduling, OpenVZ disk usage is determined by a two-level disk quota. On the first level, OpenVZ virtualization layer defines a disk quota for each VPS, for example, by limiting the maximum size of folder in the host file system. On the second level, it is possible to define disk quotas for users and groups in a VPS, using standard Linux quota mechanisms.

Network virtualization layer isolates VPSs from each other and from the physical network [SWS 05]. The default network virtualization mechanism of OpenVZ creates a virtual network interface for a VPS and assigns an IP address to it in the host system. When a packet arrives at the host system with the destination IP address of a VPS, the host system routes the packet to the corresponding VPS. This approach of network virtualization allows VPS packets to be received and sent using the host system routing module. This simplifies network virtualization, but introduces an additional hop in the route packets follow.

To conclude, OpenVZ provides a high-level virtualization abstraction and introduces less overhead than other virtualization tools. On the other hand, it is more restrictive in terms of the virtual environments that have to share the physical host system kernel.

1.3. Scenario and methodology

This section presents the methodology chosen for our tests and describes the experiments we have performed to compare the different virtualization tools. To provide a fair comparison, we decided to run different tests to evaluate CPU, RAM, storage and networking performance of the virtualized environments. To measure the performance, we chose the time taken for each tool to perform a specific task.

When virtualization is implemented, several issues arise from the fact that hardware is being divided between VMs, as previously discussed in section 1.1. One of the problems specifically related to performance measuring is how the hypervisor provides the timekeeping mechanisms to the VM [VMW 08]. There are several ways to keep track of time in a personal computer, such as reading Basic Input Output System (BIOS) clock, using CPU registers such as the Time Stamp Counter (TSC), requesting system time from the OS. For the system timekeeping mechanism, Xen constantly synchronizes the VM clock with dom0 clock by sending correct time information through shared memory between domU and the hypervisor. That solution works for most applications. Nevertheless, there will be time measurement errors if the applications have a clock frequency higher than the VM clock. For this reason, we do not consider timekeeping from the VM in our tests. The main idea of the performed tests is to have an external computer responsible for measuring how long it takes for the VMs to perform a specific task in the test computer. In the single VM tests, the external computer runs a control script, commanding the VM through an Secure Shell (SSH) connection to start the experiment and do a specific task several times. Each time a round of the task is initiated and completed, the VM notifies the external computer using a probe packet. After the completion of all the rounds, the external computer calculates the mean value and variance of the time taken to complete a round of the specified task, considering a 95% confidence interval. The number of rounds was chosen in order to have small variance and it is task dependent. In the multiple VM tests, the procedure is similar, the difference being that the external computer tracks multiple tasks executions at the same time. To make the amount of time taken to notify the external machine negligible, the tasks of all the tests were configured to make the rounds of the test last a much greater amount of time. Typical notification delays are in the order of milliseconds whereas round executions are in the order of minutes. Figure 1.7 illustrates the basic scenario of the single VM tests considering a test with only one round. The start test (1) command is sent from the external computer to the tests computer. The tests computer sends an initial notification (2) to the external computer before the task execution (3) and sends a completion notification (4) when task execution is done. Finally, the external computer processes the results (5).

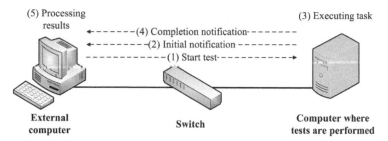

(5) Processing results

(3) Executing task

(4) Completion notification

(2) Initial notification

(1) Start test

External computer   **Switch**   **Computer where tests are performed**

Figure 1.7. *Single VM tests default configuration*

For the network tests, a different approach was taken. The metric of interest is the throughput achieved by the VMs both when receiving and sending. For these tests, it is necessary to have at least two computers, one for generating/receiving network traffic and another one for performing the complementary action. An extra computer was used for performing the complementary action of the virtualized system. This computer is referred to as Network Test Computer. For these tests, measurements were made by the network benchmarking tool itself and the used scheme is shown in Figure 1.8. The external computer initially triggers the test (1). After receiving the command, the network benchmark is executed and test packets go through the network (2). After the test completion, both the test's computer and the network test computer send the results to the external computer (3) to process the results (4).

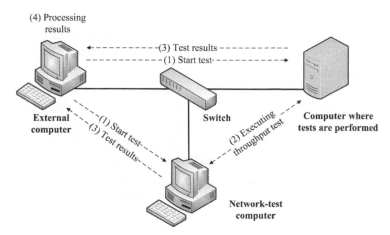

(4) Processing results

(3) Test results

(1) Start test

External computer   **Switch**   **Computer where tests are performed**

(1) Start test
(3) Test results
(2) Executing throughput test

Network-test computer

Figure 1.8. *Single VM network tests configuration*

1.3.1. *Hardware/software description*

For the performed tests, as described in section 1.3, we used three computers. In this section, we will describe the hardware and software configurations of these computers.

The tests are executed in the first computer, referred to as test's computer in Figures 1.7 and 1.8. For that role, we used an HP Proliant DL380 Generation 5 server equipped with 2 Intel Xeon 5440 CPUs, 10 GB of DDR2 RAM @667MHz, an integrated 2-port Broadcom Nextreme II Gigabit network interface, and four 146 GB SAS hard drives @10k rpm. Each hard drive was installed with a different software configuration for our tests. In the first hard drive, a standard Debian Linux AMD 64 install was made for usage in the native performance tests. In the second hard drive, a standard Xen 3.2-1 installation was made using a paravirtualized Debian Linux AMD 64 for dom0 and Domain U. The domain-U VMs were configured with one vCPU, 4 GB of RAM and 10 GB of storage. In the third hard drive, a standard VMware ESX 3.5 installation was made and the VM was configured with one vCPU, 4 GB of RAM, 10 GB of storage and Debian Linux AMD 64. In the fourth hard drive, a standard OpenVZ over Debian AMD 64 installation was made and an OpenVZ container was created with quotas similar to the Xen and VMware VMs. The second machine is the external computer used for controlling the experiments and processing the results. It is referred to as external computer in the figures. For that role, we used a desktop equipped with an Intel Q6600 CPU, 4 GB of DDR2 RAM @667 MHz, an integrated Intel e1000e Gigabit network interface and a 500 GB SATA2 hard drive @ 7k2 rpm. The computer was configured with a default Debian Linux AMD64 install and has all the scripts that control the tasks and process the results. The last computer is the computer used in network tests for establishing a connection to or receiving a connection from the tests computer. It is referred to as Network Test Computer in the figures. This computer is a desktop equipped with an Intel E6750 CPU, 2 GB of DDR2 RAM @667MHz, an integrated Intel e1000 Gigabit network interface and a 320 GB SATA2 hard drive @ 7k2 rpm. The computer was configured with a default Debian Linux AMD64 install and with the network benchmarking tool. All of the Linux installations use the 2.6.26 kernel version.

1.4. Performance evaluation

We first present the results of the experiments with one VM, which compare the performance obtained with Xen, VMware and OpenVZ against native Linux. We use specific benchmarks for CPU, RAM, hard disk and network resources. We also show the results for a Linux kernel compilation benchmark, an application that demands a mix of CPU, RAM and hard disk resources.

1.4.1. *CPU Performance*

The Super-Pi test is a CPU-intensive task based on the Gauss–Legendre algorithm to compute the Pi value. The Gauss-Legendre algorithm is iterative and is based on many arithmetic operations. The most used arithmetic operations are sum, division, square root, potentiation, subtraction and multiplication. For this test, a shell script computes the Pi value with 2^{22} digits (4,194,304 digits) for ten rounds. The performance metric is the time taken to compute Pi in seconds.

1.4.2. *Memory performance*

Memory Allocation, Set and Read (MASR) is a memory benchmarking tool developed by the Teleinformatic and Automation Group (GTA) from the Federal University of Rio de Janeiro. MASR benchmarks memory by allocating 2 GB of memory, sequentially setting all memory positions to a fixed value and sequentially reading all the memory positions. MASR was developed for benchmarking memory with a deterministic number of operations, independent of the performance of the computer. Since no Linux memory benchmarking tools were found with this explicit characteristic, MASR was developed. For this test, a shell script was developed in which MASR is executed for 10 rounds. The evaluated parameter is the time taken to execute MASR benchmark in each round.

1.4.3. *Hard disk and file system performance*

1.4.3.1. *Bonnie++*

Bonnie++ is an open-source disk benchmark designed to evaluate hard drive and file system performance. The main program of Bonnie++ tests a single temporary file of 1 GB, or multiple temporary files of 1 GB for greater amounts of data, with database-type access, simulating operations such as creating, reading and deleting large amounts of small files in the used temporary files. The second program tests the performance of the different regions of the hard drive, reading data from blocks located at the beginning, in the middle and at the end sectors of the hard drive. For this test, a shell script executes Bonnie++ for 10 rounds with a parameter for using 2 GB of disk space. The performance metric is the time taken to execute Bonnie++ benchmark each time.

1.4.3.2. ZFG

Zero File Generator (ZFG) is a disk benchmarking tool developed by GTA/UFRJ. ZFG benchmarks disk continuous writing speed by writing a 2 GB binary file filled with zeros 10 times in each round. ZFG was developed for benchmarking disk with a deterministic number of operations, independent of the performance of the computer. Since no Linux disk benchmarking tools were found with this explicit characteristic, ZFG was developed. For this test, a shell script was developed in which ZFG is executed for 10 rounds. The evaluated parameter is the time taken to execute ZFG benchmark each round.

1.4.4. *Network performance*

Iperf is an open-source network benchmark that allows benchmarking networks using both uni- and bidirectional data flows over Transmission Control Protocol (TCP) or User Datagram Protocol (UDP). Iperf has several parameters that can be configured, such as packet size, bandwidth that Iperf should try to achieve and test duration. In our test, we used Iperf configured with unidirectional UDP data flows using 1,472 bytes of payload to avoid IP fragmentation. For this test, a shell script was developed in which a unidirectional UDP data flow goes either from the test computer to the network test computer or the opposite direction.

1.4.5. *Overall performance – linux kernel compilation*

Linux kernel compilation is a benchmarking tool [WRI 02] frequently used to evaluate overall performance because it intensively uses different parts of the computer system. The Linux kernel is composed of thousands of small source-code files. Its compilation demands intensive CPU usage, RAM read/write accesses and short-duration non-sequential disk accesses, since the files to be compiled depend on the target kernel configuration. For this test, a shell script was developed in which the kernel is compiled 10 times. The evaluated parameter is the time taken to make a default configuration file and compile the kernel.

1.4.6. *Single virtual machine tests*

First, we present the results of the executed benchmarks for the single VM tests we conducted. All the following graphs regarding single VM tests will show the performance of native Linux, VMware VM, Xen VM and OpenVZ container.

1.4.6.1. *Super-Pi*

The results for the Super-Pi benchmark executed for 10 rounds are shown in Figure 1.9. The mean execution time for a round of the test is shown on the vertical axis, smaller values are better. The virtualization tools and native Linux are spread along the horizontal axis.

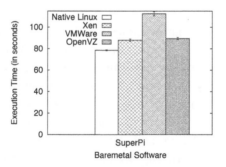

Figure 1.9. *Super-Pi test*

As expected, the non-virtualized system outperforms the virtualized systems. Xen is close to native performance. That was expected since Xen paravirtualized VMs can directly access the memory after hypervisor approval for using a memory area and we associate most of the presented overhead to the VM CPU scheduler. VMware has the worst performance, a consequence of the extra address translation between the virtual RAM offered to the VM and the physical RAM, and also to the VM CPU scheduler. OpenVZ has an overhead a little higher than Xen, due to the scheduling mechanism sharing CPU between the containers.

1.4.6.2. *MASR*

The results for the MASR benchmark executed for 10 rounds are shown in Figure 1.10. The mean execution time for a round of the test is shown on the vertical axis, smaller values are better. The virtualization tools and native Linux are spread along the horizontal axis.

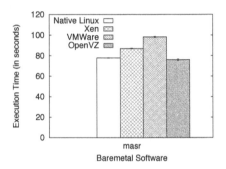

Figure 1.10. *MASR test*

The results show that native Linux has the best performance as expected. OpenVZ achieves similar performance to native Linux that was expected since OpenVZ shares memory dynamically with all the containers with a low isolation level that only guarantees a minimal amount of private memory for each container. Xen presents some overhead associated with the VM scheduling mechanism. VMware has the worst performance due to the extra memory address translation and VM scheduler. All virtualization tools presented acceptable overhead, which is a very important result, since memory access operations are very common in forwarding table lookup and other routing tasks.

1.4.6.3. *Bonnie++*

The results for the Bonnie++ disk benchmark executed for 10 rounds are shown in Figure 1.11. The mean execution time for a round of the test is shown on the vertical axis, smaller values are better. The virtualization tools and native Linux are spread along the horizontal axis.

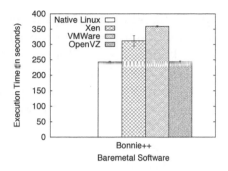

Figure 1.11. *Bonnie++ test*

Native Linux has the best performance and OpenVZ achieves similar results since disk access in OpenVZ is very similar to disk access in native Linux. Xen presented some overhead because it uses an extra buffer between dom0 and the VM, the I/O ring, which adds latency when reading from disk. VMware presents the worst results, probably a consequence of the extra buffers in the hypervisor that are used to transfer data between the physical disk and the virtual disk. Nevertheless, disk performance is of minor importance since it does not impact much normal router forwarding operation.

1.4.6.4. *ZFG*

The results for the ZFG test are shown in Figure 1.12. VMware, Linux and OpenVZ results were obtained from 10 rounds. Xen results were obtained from 100 rounds in order to have an acceptable error bar. The mean execution time for a round of the test is shown on the vertical axis, smaller values are better. The virtualization tools and native Linux are spread along the horizontal axis.

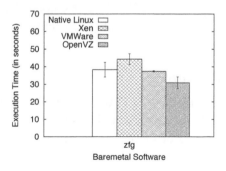

Figure 1.12. *ZFG test*

As observed in Figure 1.12, VMware VM performed a little faster than native Linux and we credit this to the buffer that transfers data between the virtual disk and the real disk, since it gives VM the impression that data has already been written to disk when it is actually being transferred from the buffer to the physical disk. OpenVZ unexpectedly outperformed native Linux and further investigation must be performed to discover the reason. Xen VM performed slower than native Linux but an irregular execution time was observed. In Figure 1.13, we note that several rounds took less than 10 s to execute and a few took more than 60 s to execute. In native Linux, when there are data to be written in an I/O device, it is sent to a memory buffer and then it is transferred to the device using Direct Memory Access (DMA). In Xen and VMware, the extra step in which the virtual I/O device driver writes the data to a memory buffer used to communicate with the real I/O device driver gives the impression that data were written in the hard drive, when it is actually in the buffer from the VM do the real I/O device driver. Further details on this buffer for the Xen implementation can be found in [CHI 08]. The great round execution time difference presented by Xen results corroborate our hypothesis since the time needed for each round would vary according to the state of the buffer in the beginning of each round. To confirm our suspicion, we made two kinds of further test. The first kind of test is for handicapping the VMs buffer. In these kinds of test, we modified the test scripts to write large amounts of data before each round of the test in order to dirty the buffer from the VM to the real I/O device driver. In these kinds of tests, it is expected that the buffer from the VM to the real IO driver is filled with pending requests at the beginning of each round. The second kind of test favors the VMs buffer. We modified the script to wait for a long period before each round, giving dom0 enough time to clear the buffer. In these kinds of tests, it is expected that the buffer from the VM to the real I/O driver is empty, that is without pending requests, at the beginning of each round. The results of the additional test are discussed in the following.

1.4.6.5. *ZFG with buffer dirtying*

The results for the modified ZFG test in which we dirty the buffer with large amounts of data before each round is shown in Figure 1.14. The mean execution time for a round of the test is shown on the vertical axis, smaller values are better. The virtualization tools and native Linux are spread along the horizontal axis.

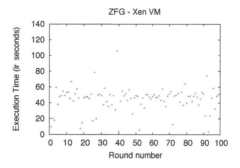

Figure 1.13. *ZFG test Xen VM rounds*

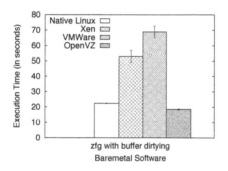

Figure 1.14. *ZFG with buffer dirtying test*

The test was made with 100 rounds and, as expected, when we dirty the buffer that transfers data between the virtual and real I/O device driver, the VMs performance drops significantly and native Linux outperforms both Xen and VMware virtualized Linux OSs with great advantage. Again, OpenVZ outperforms native Linux, indicating that the disk mechanism adopted by OpenVZ is not similar to the disk mechanism adopted by Xen and VMware VMs.

1.4.6.6. *ZFG with resting period*

The results for the modified ZFG test in which we wait dom0 to clear the buffers before each round are shown in Figure 1.15. The mean execution time for a round of the test is shown on the vertical axis, smaller values are better. The virtualization tools and native Linux are spread along the horizontal axis.

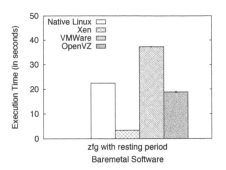

Figure 1.15. *ZFG with resting period test*

As expected, when there is a long period of time between the rounds, Xen has enough time to write all the pending data that is in the buffer and when the next round of the test is executed, the buffer is clear, giving the impression that Xen VM writes data to the disk faster. Therefore, when the amount of written data is not much bigger than the buffer, Xen VM has such good performance because it is writing on the RAM memory buffer that connects the virtual disk driver and the real disk driver, outperforming native Linux that is actually writing data onto the hard drive. We allowed 2 min between each round for Xen to clear the buffers. There is no such mechanism in VMware that did not perform as well. OpenVZ once more outperformed Linux as in the previous ZFG disk writing tests.

1.4.6.7. *Iperf: UDP data flow from VM to network test computer*

In this test, UDP packet generation from the VM is being benchmarked. Results for packet reception are shown in the next section. The results for the Iperf network benchmark generating traffic from the VM to the Network Test Computer for 100 rounds are shown in Figure 1.16. The mean throughput is shown on the vertical axis, greater values are better. The virtualization tools and native Linux are spread along the horizontal axis.

As expected, native Linux achieves the best results, near the theoretical limit of the gigabit Ethernet adapter. Xen achieves near-native performance showing that the I/O ring mechanism is capable of delivering data fast enough. VMware presents great performance degradation that points out an inefficient implementation of its default networking mechanism. OpenVZ default networking mechanism based on IP layer virtualization performed poorly and could not use the full gigabit interface bandwidth. This is a very important result since the forwarding capability of a router directly depends on its capability of sending packets, and the results show that it is possible for a virtualized environment to send large packets at near-native speed.

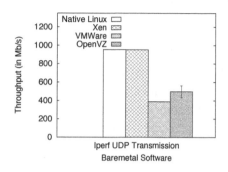

Figure 1.16. *Iperf UDP packet generation test*

1.4.6.8. Iperf: UDP data flow from network test computer to VM

In this test, UDP packet reception from the VM is being benchmarked. The results for the Iperf network benchmark generating traffic from the Network Test Computer to the VM for 100 rounds are shown in Figure 1.17. The mean throughput is shown on the vertical axis, greater values are better. The virtualization tools and native Linux are spread along the horizontal axis.

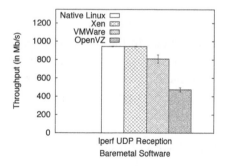

Figure 1.17. *Iperf UDP test*

The results show once again that native Linux is close to the theoretical maximum throughput. Xen achieves near-native performance once again. VMware performed much better in reception than in transmission but still worse than native Linux and Xen. The default networking virtualization mechanism of OpenVZ is performed poorly. This is also an important result since the forwarding capability of a router is directly related to its capability of receiving packets. The results show that Xen can handle large packets reception in near-native speed with a single VM, multiple VM tests are presented in section 1.4.7.

1.4.6.9. *Linux kernel compilation*

The results for the Linux kernel compilation test executed for 10 rounds are shown in Figure 1.18. The mean execution time for a round of the test is shown on the vertical axis, smaller values are better. The virtualization tools and native Linux are spread along the horizontal axis.

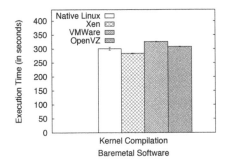

Figure 1.18. *Kernel compilation test*

Once again Xen achieves a slightly better result than native Linux, due to the disk writing effect explained in the previous disk writing tests. VMware achieves the worst results for this task repeating the poor CPU and memory performance presented in the previous tests. OpenVZ performance is similar to native Linux as expected by the performance presented in the previous CPU, memory and Bonnie++ disk tests.

1.4.6.10. *Remarks on the single VM results*

An important point to comment on is the disk performance of VMs. The buffer that communicates the real disk driver and the virtual disk driver can have either a positive or negative effect depending on the application. For applications that make numerous disk reading operations, the extra stage between the hard drive and the VM disk driver introduces longer delays for recovering information, therefore decreasing performance. If disk writing operations are more frequent than reading and the application does not write the large amounts of data to fill the buffer, then the application should benefit from the buffer, just like the ZFG benchmark when executed with an interval between the rounds. Applications that would lose performance include website hosts for sites that have to frequently recover information from the disk or data mining applications that intensively read information from the disk. Applications that would benefit from the buffer include logging applications and other applications that mostly access disk for writing data. Considering the presented results, especially the CPU, memory access and network related, we conclude that, at first, virtualization tools introduce acceptable overhead, and building virtual routers over the tested tools is feasible. Based on the CPU, memory and network results, which are the main resources used in router normal operation, we conclude that VMware is not the best option for building virtual routers since it had the worst performance results. Also, VMware provides little flexibility since its source code is proprietary. OpenVZ performed very well in most of the tests, but presented poor network performance, which is a major issue for network virtualization. Moreover, OpenVZ also restricts the virtual environments with the limitation of using, not only the same OS, but also the same kernel version for all VMs, which is a major issue for building virtual networks with different levels of router complexity. For presenting good CPU and memory performance, having near-native network performance, being open source allowing customization with no restrictions and for providing virtual environments with a high degree of flexibility, Xen is probably the best virtualization tool for building virtual routers. Section 1.4.7 presents the set of tests with multiple VMs. The multiple VM tests evaluate how Xen scales with multiple VMs sharing the same computer resources, assessing the viability of multiple virtual networks over a computer.

1.4.7. *Multiple virtual machine tests*

The objective of the multiple VM tests is to observe how Xen scales with multiple VMs regarding performance degradation, fair resource sharing and how different vCPU to CPU allocation schemes affect overall and individual performance. All the following graphs show the performance of Xen VMs, varying the number of VMs performing different tasks and varying the CPU allocation scheme for the same task. Xen allows configuring how the vCPU presented from the VMs to the virtualized OSs share the physical CPUs. It allows both dynamic CPU allocation schemes and static allocation schemes.

Three different CPU allocation schemes were used in order to study their influence in the scalability of the virtualized environments. As previously mentioned, the test computer is equipped with two Xeon E5400 quad-core CPUs that give eight physical cores to distribute between the hypervisor and VMs' needs. Another relevant point is that each CPU has its own front side bus, allowing each CPU to access memory independently. Inside each CPU, each core has its own L1 and L2 caches, but the L3 cache is shared for each pair of cores. The first core allocation scheme used in the tests is Xen Default scheme, in which the hypervisor dynamically allocates the CPU physical cores to the Xen vCPUs. In Xen Default scheme, there can be unused physical cores, cores being allocated to more than one vCPU and cores being allocated to exactly one vCPU. An example of core allocation is presented in Table 1.1. At a certain moment, there can be unused physical cores, cores allocated to more than one vCPU and cores allocated to a single vCPU.

Physical core	Virtual CPU of virtual machine
0	VCPU0 and VCPU7 of dom0
1	VCPU4 of dom0
2	VCPU1 of dom0
3	VCPU0 of VM2
4	VCPU5 and VCPU6 of dom0; VCPU0 of VM4
5	VCPU2 of dom0; VCPU0 of VM3
6	Not used
7	VCPU3 of dom0; VCPU0 of VM1

Table 1.1. *Example of the physical core allocation using Xen Default scheme*

The second scheme we used is a static physical core allocation scheme. We set a fixed exclusive core to each VM vCPU and a fixed exclusive core to four of dom0 vCPUs, disabling four of its vCPUs. The objective of this allocation is to observe if there is a considerable gain of performance when fixing the physical core which will handle a vCPU request, especially leading to a reduced number of CPU cache faults, decreasing RAM memory accesses that have much greater latency than cache. The scheme is shown in Table 1.2.

The third scheme we used is a static physical core allocation in which we set all the vCPUs to share the same physical core. The objective of this scheme is to overload the physical CPU, observing how performance decreases and if resource sharing between the VMs is fair. The scheme is shown in Table 1.3.

Each VM and dom0 use the same fixed physical core. Dom0 has only one active vCPU and all the other seven vCPUs are disabled.

Physical core	Virtual CPU of virtual machine
0	VCPU0 of dom0
1	VCPU1 of dom0
2	VCPU2 of dom0
3	VCPU3 of dom0
4	VCPU0 of VM1
5	VCPU0 of VM2
6	VCPU0 of VM3
7	VCPU0 of VM4

Table 1.2. *Exclusive core static allocation scheme*

Physical core	Virtual CPU of virtual machine
0	VCPU0 of dom0; VCPU0 of VM1; VCPU0 of VM2; VCPU0 of VM3; VCPU0 of VM4
1	Not used
2	Not used
3	Not used
4	Not used
5	Not used
6	Not used
7	Not used

Table 1.3. *Shared core static allocation scheme*

1.4.7.1. *Super-Pi*

The results for the scalability tests using the Super-Pi benchmark with the three different physical cores' allocation schemes, executed for 10 rounds each, are shown in Figures 1.19a)–c).

The results show that all the three CPU schemes present much greater variance, as the number of VMs is increased showing that for this task Xen had difficulty to schedule the VMs in a fair way. For the scheme with one physical core serving all the vCPUs, it is also noticeable that there was the largest observed variance and, since CPU was a scarce resource, the more VMs were executing, the longer a Super-Pi round would take.

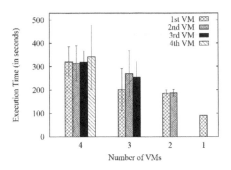

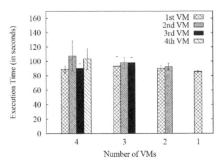

a) Multiple VM using one physical core
for all the vCPUs

b) Multiple VM using one exclusive
physical core for each vCPU

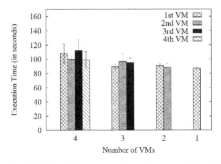

c) Multiple VM using Xen Default physical
core allocation scheme

Figure 1.19. *Super-Pi benchmark*

1.4.7.2. *MASR*

The results for the scalability tests using the MASR benchmark with three different physical cores allocation schemes executed for 10 rounds are shown in Figures 1.20a)–c).

The results show that both Xen scheme and the scheme with one exclusive physical core for each VM and for each dom0 vCPU scale well from one to four VMs, showing no degradation of performance as the VM number increases. Xen default scheme does not lose much performance when scaling from one to four VMs, but presents worse performance than the scheme with one exclusive physical core for each VM and for each active dom0 vCPU. That is a consequence of the fact that Xen default scheme reassigns vCPUs over the physical cores dynamically, increasing cache misses when a vCPU is reassigned to a different core. The scheme using one physical core for all the VMs and Dom0 shows that for this test CPU is a bottleneck and that Xen hypervisor is fair regarding the division of the CPU among the VMs.

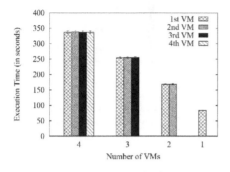

a) Multiple VM using one physical core for all the vCPUs

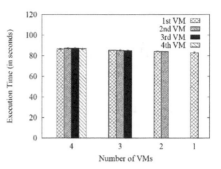

b) Multiple VM using one exclusive physical core for each VM vCPU and for each dom0 vCPU

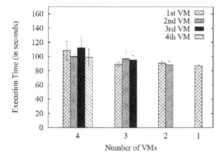

c) Multiple VM using Xen Default physical core allocation scheme

Figure 1.20. *MASR benchmark*

1.4.7.3. *ZFG with buffer dirtying*

The single VM tests have revealed that Xen VMs do not actually write data to the hard drive synchronously. Thus, we chose to execute only the disk test that was as close as possible from writing to the hard drive synchronously. In this test, the buffer is dirtied from the VM to the real hard disk driver in the beginning of each round. The results for the scalability tests using the ZFG with buffer dirtying benchmark with three different physical cores allocation schemes executed for 10 rounds are shown in Figures 1.21a)–c).

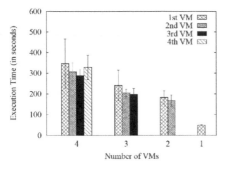

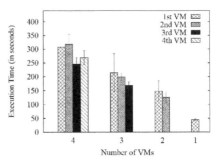

a) Multiple VM using one physical core for all the vCPUs

b) Multiple VM using one exclusive physical core for each VM vCPU and for each dom0 vCPU

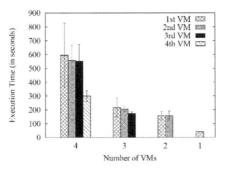

c) Multiple VM using Xen Default physical core allocation scheme

Figure 1.21. *ZFG with buffer dirtying benchmark*

The results show that all the three CPU schemes present much greater variance as the number of VMs increases, showing that for this task Xen had difficulty to schedule the VMs fairly. The scheme that presented the best performance was the scheme with one exclusive physical core for each vCPU. The scheme with one physical core serving all the vCPUs presented small overhead when compared to the one exclusive physical core for each vCPU scheme. That can be explained by the fact that in this task the bottleneck is the hard drive, even with only one core for all the vCPUs. Surprisingly, Xen Default scheme presented low performance when the task was executed with four VMs showing instability. Further investigation is required to understand why this instability was observed.

1.4.7.4. *Iperf: UDP data flow from VM to network test computer*

The results for the scalability tests using the Iperf benchmark generating UDP data flow from VM to the Network Test Computer with three different physical cores allocation schemes executed for 10 rounds are shown in Figures 1.22a)–c).

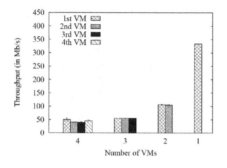

a) Multiple VM using one physical core for all the vCPUs

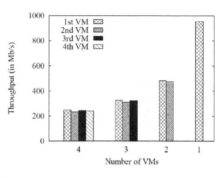

b) Multiple VM using one exclusive physical core for each VM vCPU and for each dom0 vCPU

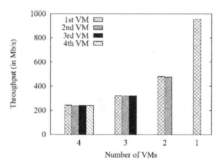

c) Multiple VM using Xen Default physical core allocation scheme

Figure 1.22. *Iperf: UDP data flow from VM to Network Test Computer*

The results show that both Xen Default physical core allocation scheme and one exclusive physical core for each VM vCPU scheme had similar performance achieving total throughput near the theoretical limit of the network adapter. For the one physical core for all vCPUs scheme, the bottleneck was not the network adapter but the CPU since the VMs were not able to generate enough packets to fill the network adapter capacity. In all schemes, Xen has proven to be fair in resource sharing.

1.4.7.5. *Linux kernel compilation*

The results for the scalability tests using the Linux kernel compilation with three different physical cores allocation schemes executed for 10 rounds are shown in Figures 1.23a)–c).

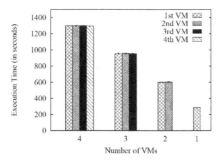

a) Multiple VM using one physical core for all the vCPUs

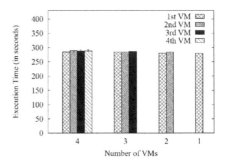

b) Multiple VM using one exclusive physical core for each VM vCPU and for each dom0 vCPU

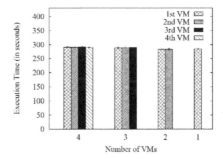

c) Multiple VM using Xen Default physical core allocation scheme

Figure 1.23. *Linux kernel compilation*

The results show that both Xen Default scheme and the scheme with one exclusive physical core for each VM and for each dom0 vCPU scale well from one to four VMs, showing no degradation of performance. The scheme using one physical core for all the VMs and dom0 has shown that for this test CPU was a bottleneck and that Xen hypervisor was fair regarding the division of the CPU among the VMs.

1.4.7.6. *Conclusions*

In general, Xen has proven to be fair when scaling to multiple VMs. For some tests, scaling multiple VMs has affected fairness, which was shown by the increasing variance in the results as the number of VMs would increase. It is also noticeable that except for the hard drive test, concentrating all the vCPUs in one physical core has considerably decreased VMs' performance. Except for the memory tests, there were no noticeable differences between the performance of Xen Default scheme and the one physical core for each VM vCPU scheme showing that CPU cache preservation was not a major issue for the tasks we executed. An unexpected result that would need extra investigation is Xen Default scheme's poor performance in the hard drive test when four VMs were executing the test.

1.5. Summary

The efforts to propose a new Internet architecture can be classified into two groups: monistics or purists, in which the proposed architecture is composed of a single general purpose network to support any service requirement, and pluralistics, in which the proposed architecture is composed of multiple networks, each one built to cope with a kind of service requirement. In this chapter, we have analyzed the role of current hardware virtualization techniques in the support of pluralistic architectures.

We have compared VMware, Xen and OpenVZ as candidates for implementing router virtualization. As discussed in section 1.4.6.10, we conclude that VMware is not the best option for router virtualization, because it achieved the worst results in the tests and is a proprietary solution that does not allow modifications since there is no access to its source code. We also concluded that, although OpenVZ had great performance results in most of the tests, it is also not the best choice since it achieved low network performance, which is one of the main concerns for building virtual routers. OpenVZ also provides low flexibility for the construction of virtual routers, because all the virtual environments must share not only the same OS but also the same kernel. Xen presented the best fit for the virtual router requirements. Xen provides flexibility for allowing any OS inside the VMs and for allowing customizations, since it is an open-source tool. Xen has also shown good CPU, memory and virtualized network performance, which are the most important resources for virtual routers' forwarding operation. Moreover, Xen provides fair resource sharing between the VMs and has scaled from one to four VMs with no performance degradation caused by the hypervisor.

1.6. Bibliography

[ADA 06] ADAMS K., AGESEN O., "A comparison of software and hardware techniques for x86 virtualization", *ASPLOS-XII: Proceedings of the 12th International Conference on Architectural Support for Programming Languages and Operating Systems*, 2006.

[BAR 03] BARHAM P., DRAGOVIC B., FRASER K., *et al.*, "Xen and the art of virtualization", *Proceedings of the 19th ACM Symposium on Operating Systems Principles – SOSP03*, October 2003.

[CHI 08] CHISNALL D., *The Definitive Guide to the Xen Hypervisor*, Prentice Hall, 2008.

[EGI 07] EGI N., GREENHALGH A., HANDLEY M., *et al.*, "Evaluating Xen for router virtualization", *International Conference on Computer Communications and Networks – ICCCN*, pp. 1256–1261, August 2007.

[KOL 06] KOLYSHKIN K., *Virtualization in Linux*, 2006.

[MEN 06] MENON A., COX A.L., ZWAENEPOEL W., "Optimizing network virtualization in Xen", *USENIX Annual Technical Conference*, pp. 15–28, May 2006.

[POP 74] POPEK G.J., GOLDBERG R.P., "Formal requirements for virtualizable third generation architectures", *Communications of the ACM*, vol. 17, no. 7, pp. 412–421, 1974.

[SWS 05] SWsoft Inc., *OpenVZ Users Guide*, 2005.

[VMW 05] VMWare Inc., *VMware ESX Server 2 Architecture and Performance Implications*, 2005.

[VMW 07a] VMWare Inc., *The Architecture of VMware ESX Server 3i*, 2007.

[VMW 07b] VMWare Inc., *A Performance Comparison of Hypervisors*, 2007.

[VMW 08] VMWare Inc., *Timekeeping in VMware Virtual Machines*, 2008.

[WRI 02] WRIGHT C., COWAN C., MORRIS J., *et al.*, "Linux security modules: general security support for the linux kernel," *Proceedings of the 11th USENIX Security Symposium*, August 2002.

[XEN 07] XenSource, Inc., *A Performance Comparison of Commercial Hypervisors*, 2007.

Chapter 2

Virtual Network Interfaces

Future Internet architecture proposals can be divided into two main approaches: monist and pluralist. In the monist model, shown in Figure 2.1a), the network has a monolithic architecture that must be flexible enough to support all sorts of upcoming applications. In this approach, only one protocol stack runs over the physical substrate at a time. On the other hand, the pluralist approach, shown in Figure 2.1b) [AND 05], is based on the idea that the Internet must simultaneously support multiple networks, each network running a protocol stack that fits the needs of a given application. Creating specialized networks designed for specific services simplifies the deployment of new applications requiring features such as security, mobility or Quality-of-Service. We claim that designing multiple networks to provide different services is easier than designing a unique network to handle all the different services at the same time. The pluralist approach, therefore, can be understood as an umbrella based on the "divide-and-conquer" approach. If it is very difficult to find a single solution to cover all possible requirements, a viable alternative can be created with multiple different networks to support them. This characteristic is the main argument for the pluralist architecture, because the monist approach not only has to solve all the known Internet problems, but it must also evolve to be continuously operational in the future. Furthermore, another key advantage of the pluralist model is its intrinsic

Chapter written by Miguel Elias M. Campista.

backward compatibility, since the current Internet can be one of the supported networks running in parallel.

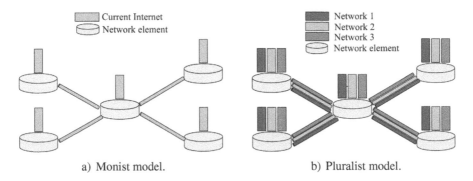

a) Monist model.   b) Pluralist model.

Figure 2.1. *Network architectures: monist model with only one protocol stack and pluralist model with several protocol stacks*

Pluralist proposals are based on the same idea of multiple networks running over the same infrastructure [MAC 11, FEA 07, VER 07], even though they may differ in packet formats, addressing schemes, and protocols. Because these networks share the same physical network substrate, for example routers and links, the multiple accesses to the underlying medium must be orchestrated by a monitor. This monitor is a piece of specific-purpose software virtualizing the shared medium to the multiple networks running atop. Hence, each network runs as if it was using a given amount of physical resources. These networks are called virtual networks and must deal with performance challenges as a consequence of using a shared physical substrate. This additional task is the overhead of the pluralist architecture compared with the monist architecture because it adds a software layer.

Based on the pluralist approach, previous works, such as the works developed by the Horizon Project team [HOR 10], propose the use of two different virtualization platforms to provide virtual networking: Xen and OpenFlow. These platforms tackle, in a different fashion, the problem of physical resource sharing among multiple virtual networks. The differences lead to a trade-off between performance and flexibility, as will be further discussed.

In this chapter, section 2.1 first presents the two virtualization platforms: Xen and OpenFlow. Then, it provides a performance analysis focusing on the main characteristics of each platform. Afterwards, we describe interfaces for system management, which must be friendly enough to allow simple resource sharing management among multiple virtual networks and must also facilitate actions upon them, such as virtual network topologies reconfiguration. The Horizon Project designed and developed interfaces for both Xen and OpenFlow platforms such as VNext [PIS 11] and OMNI [MAT 11], respectively, and built a prototype for each platform for the sake of proof of concept [HOR 10]. Most of the developed interfaces do not require human intervention and can also be used by computer agents to autonomously control the network. Consequently, these interfaces can be used for advanced piloting systems that are able to provide autonomous networking services. Section 2.2 describes in more detail the interfaces used in the Xen prototype, whereas section 2.3 describes the interfaces of the OpenFlow prototype. Finally, section 2.4 concludes this chapter.

2.1. Virtual networks: isolation, performance and trends

This section describes the issue of sharing the network physical substrate among different virtual networks[1]. We show the analysis of two representative approaches for virtualizing physical networks, Xen [EGI 07] and OpenFlow [MCK 08], and discuss the use of these technologies for virtual networking. The main goal is to investigate the pros and cons of network virtualization, used as the basis of the pluralist architecture for the future Internet. To this end, experiments are conducted to evaluate Xen and OpenFlow [MAT 09] performance acting as a virtualized software router. Based on the presented results and on previous work, we have discovered a trade-off between flexibility and performance, indicating that the use of shared data planes could be an important architectural choice when using virtual networks. Another key point is that, using shared data planes, Xen and OpenFlow can multiplex several virtual networks without any measurable performance loss, compared with a scenario in which the same packet rate is handled by a single virtual network element.

2.1.1. *Network virtualization approaches*

We consider virtualization as a resource abstraction that allows the slicing of a physical resource into several virtual resources of the same type but with

---

1 This section partially appeared in our previous work [FER 11].

different capacities, as shown in Figure 2.2. This abstraction is often implemented by a software layer that provides "virtual sliced interfaces" quite similar to the real interfaces. The coexistence of several virtual slices over the same resource is possible because the virtualization layer decouples the real resource and the above layer. Figure 2.2 shows two examples of virtualization: computer virtualization and network virtualization. The computer virtualization abstraction is implemented by the virtual machine monitor (VMM), which provides virtual machines (VMs) with an interface (i.e. the hardware abstraction layer) quite similar to a computer hardware interface. This interface includes a processor, memory, input/output (I/O) devices, etc. Thus, each VM has the impression of running directly over the physical hardware, but actually the physical hardware is shared among several VMs. We call this kind of resource sharing slicing, because the VMs are isolated: one VM cannot interfere with another. Computer virtualization is widely used in datacenters to run several servers in a single physical machine. Besides saving energy and reducing maintenance costs, this technique has, as its most important feature, flexibility, yielding each VM to have its own operating system, applications, configuration rules, and administrative procedures. The possibility to run a customized protocol stack into each virtual slice is the main motivation of applying virtualization to networks [AND 05]. As shown in Figure 2.2, network virtualization is analogous to computer virtualization, in which the shared resource is the network instead. This concept is not new and it has been used in virtual private networks (VPNs) and virtual local area networks (VLANs). Nowadays, however, there are new techniques allowing even router virtualization. In this case, each virtual router slice can implement a customized network protocol stack.

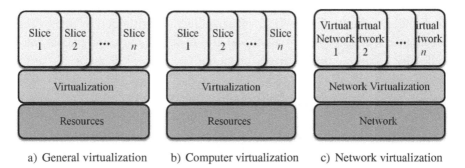

a) General virtualization   b) Computer virtualization   c) Network virtualization

Figure 2.2. *Virtualization: "sliced" resources, virtual machines and virtual routers*

Approaches for network virtualization contrast on the level of abstraction they provide, which can be illustrated by the position of the virtualization layer in the architectural design. Figure 2.3 compares two basic approaches for virtualizing a network element. Figure 2.3a) shows the conventional network element architecture with a single control and data plane. In a router, the control plane is responsible for running the network control software, such as routing protocols (e.g. Routing Information Protocol (RIP), Open Shortest Path First (OSPF) and Border Gateway Protocol (BGP)) and network control protocols (e.g. Internet Control Message Protocol (ICMP)), whereas the data plane is responsible for the implementation of forwarding tables and hardware datapaths. Virtualizing the routing procedure means inserting a virtualization layer at some level of the network element architecture to allow the coexistence of multiple virtual network elements over a single physical network element. Assuming there is a virtualization layer between control and data planes, then only the control plane is virtualized, as shown in Figure 2.3b). In this case, the data plane is shared by all virtual networks and each virtual network runs its own control software. Compared with the conventional network architecture, this approach greatly improves network programmability because it allows running multiple and customized protocol stacks, instead of a single default protocol stack. For instance, it is possible to program different protocol stacks for network 1, network 2 and network 3, as illustrated in Figure 2.3b). In the second network virtualization approach, both control and data planes are virtualized, as shown in Figure 2.3c). In this case, each virtual network element implements its own data plane, beside the control plane, improving even more the network programmability. This approach allows data plane customization at the cost of packet forwarding performance because the data plane is no longer dedicated to a common task. This trade-off between network programmability and performance is discussed in detail in sections 2.1.3.1 and 2.1.3.2.

It is worth mentioning that by only virtualizing the control plane, it is possible to divide the virtualization approach into more categories depending on the isolation level of the data plane shared among the virtual network elements. If a strong isolation is required, then each virtual control plane must only have access to its slice of the data plane and cannot interfere on the others. On the other hand, if the whole data plane is shared among virtual control planes, then it is possible to have a virtual control plane interfering on other virtual control planes. For instance, a single virtual control plane can fill the entire forwarding table with its own entries, which can lead other virtual networks to have their packets dropped.

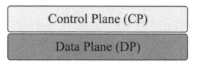

a) Conventional model

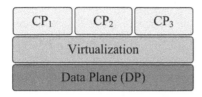

b) Pluralist model with only virtualized control plane (CP)

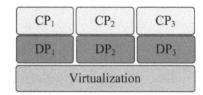

c) Pluralist model with virtualized control (CP) and data planes (DP)

Figure 2.3. *Network models*

2.1.2. *Network virtualization technologies*

In this section, we present, in detail, two technologies that can be used for network virtualization: *Xen* and *OpenFlow*.

Xen is an open-source VMM, also called hypervisor, that runs on commodity hardware platforms [EGI 07]. Xen architecture is composed of one VMM, located above the physical hardware, and several domains, running simultaneously above the hypervisor, called VMs, as shown in Figure 2.4. Each VM can have its own operating system and applications. The VMM controls the access of the multiple domains to the hardware and also manages the resource sharing among these domains. Hence, one of the main tasks of the VMM is to isolate the different VMs, i.e. the execution of one VM must not affect the performance of the others. In addition, all the device drivers are kept in an isolated driver domain, called *Domain 0* (dom0), in order to provide reliable and efficient hardware support [EGI 07]. Domain 0 has special privileges compared with the other domains, referred to as *unprivileged domains* (domUs), because it has total access to the physical machine hardware. Unprivileged domains, also called user domains, have virtual drivers and operate as if they could directly access the hardware. Nevertheless, these virtual drivers communicate with dom0 to have access to the physical hardware.

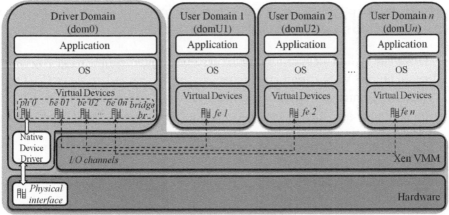

fe – front-end interface   be – back-end interface   ph – physical interface

Figure 2.4. *Xen architecture: virtual machine monitor (VMM), driver domain (dom0) and user domains (domUs)*

Xen virtualizes a single physical network interface by demultiplexing incoming packets from the physical interface to the domUs and, conversely, multiplexing outgoing packets generated by these domUs. This procedure, called network I/O virtualization, works as follows. Domain 0 directly accesses I/O devices by using its native device drivers and it also performs I/O operations on behalf of domUs. On the other hand, domUs employ virtual I/O devices, controlled by virtual drivers, to request dom0 for device access [MEN 06], as illustrated in Figure 2.4. Each domU has its own virtual network interfaces, called front-end interfaces, required for network communications. Back-end interfaces are created in dom0 corresponding to each front-end interface in a domU and act as a proxy for the virtual interfaces in dom0. The front-end and back-end interfaces are connected to each other through an I/O channel, which employs a zero-copy mechanism to remap the physical page containing the packet into the target domain. In this way, packets are exchanged between the back-end and the front-end interfaces [MEN 06]. The front-end interfaces are perceived by the operating systems running on domUs as the real interfaces. Nevertheless, the back-end interfaces in dom0 are connected to the physical interface and also to each other through a virtual network bridge. This is the default architecture used by Xen and it is called the bridge mode. Thus, both the I/O channel and the network bridge establish a communication path between the virtual interfaces created in domUs and the physical interface.

Different virtual network elements can be implemented using Xen as it allows multiple VMs to run simultaneously on the same hardware [EGI 07], as shown in Figure 2.5. In this case, each VM runs a virtual router, which has its own control and data planes because the Xen virtualization layer is at a low level.

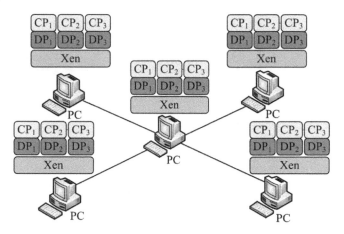

Figure 2.5. *Xen virtual networking: one data and control planes per virtual router*

OpenFlow [MCK 08] allows the use of the wiring closets on university campus not only for the production network, but also for experimental networks. The OpenFlow project, proposed by Stanford University, aims at creating virtual environments for innovations in parallel with the production network using network elements such as switches, routers, access points and personal computers.

OpenFlow presents a new architecture for providing virtual network environments. The key idea is the physical separation of control and data planes. Therefore, different network elements execute the packet forwarding procedure (data plane) and the network control procedure (control plane). The virtualization of the forwarding elements is accomplished by a shared flow table, which represents the data plane, whereas all the control planes are centralized in a node called controller, running applications that control each virtual network. An example of OpenFlow operation is shown in Figure 2.6.

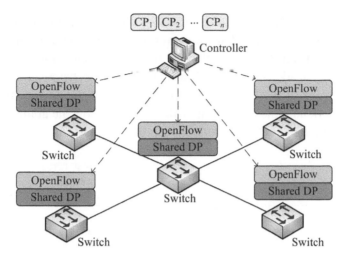

Figure 2.6. *OpenFlow virtual networking: a shared data plane per node and all the control planes in one node*

The OpenFlow protocol defines the communication between forwarding nodes and the network controller. It is based on the establishment of a secure channel between each forwarding node and the controller, which uses this channel to monitor and configure the forwarding nodes. Every time a new packet reaches a forwarding element and there is no previously configured flow, the packet's first bits are forwarded to the controller, which sets a path for this packet in the chosen forwarding elements. The controller may also set the action of normal processing for a flow to be forwarded according to conventional layer-2 (L2) and layer-3 (L3) routing, as if OpenFlow did not exist. That is the reason why OpenFlow can be used in parallel to the production network without affecting the ongoing traffic.

The data plane in OpenFlow is a flow table described by header fields, counters and actions. The header fields are a 12-tuple structure that describes the packet header, as shown in Figure 2.7. These fields specify a flow by setting a value for each field or by using a wildcard to set only a subset of fields. The flow table also supports the use of subnet masks, if the hardware in use also supports this match [PFA 09]. This 12-tuple structure gives high flexibility for packet forwarding because a flow can be forwarded based not only on the destination Internet Protocol (IP) address, as in the conventional Transmission Control Protocol (TCP)/IP network, but also on the TCP port, the Medium Access Control (MAC) address, etc. Because the flows can be set based on layer-2 addresses, the forwarding elements of OpenFlow are also called OpenFlow switches. This, however, does not necessarily imply packet forwarding in layer-2. One of the future goals of OpenFlow is to handle user-described header fields, which means that the packet header will not be described by fixed fields, but by a combination of fields specified by the virtual network administrators. This will give OpenFlow the ability to forward packets belonging to networks running any protocol stack.

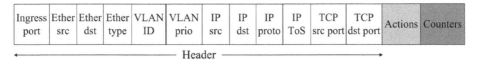

Figure 2.7. *A flow entry in an OpenFlow network*

After the header fields, the flow description is followed by counters, which are used for node monitoring. Counters compute data such as the flow duration and the amount of bytes forwarded. The last fields in the flow description are the actions, which are a set of instructions that can be taken over each packet of a specific flow in the forwarding elements. These actions include not only switching an incoming packet to a destination port, but also changing header fields such as the VLAN identifier and source and destination addresses.

The controller node is a central element in the network and can communicate with all nodes to configure their flow tables. The controller runs a network operating system, which provides virtual network management applications, the basic functions for network configuration. Hence, the controller in OpenFlow works as an interface between the network applications and the forwarding elements, providing the basic functions to access packets from a flow and to monitor nodes. OpenFlow works with any controller that is compatible with the OpenFlow protocol, such as network operating system (NOX) [GUD 08]. In this case, each control plane is composed of a set of applications running over NOX. Hence, a virtual network in OpenFlow is defined by its control plane, which is a set of applications running over the controller, and its respective flows under control, as shown in Figure 2.8.

Using the single controller model, it is possible to create many virtual networks. Note, however, that different applications running over the same operating system are not isolated. As a result, if one application has some bug, it can stop the controller, harming all other virtual networks. FlowVisor is a tool used with OpenFlow to allow different controllers to work over the same physical network [SHE 10]. FlowVisor works as a proxy between the forwarding elements and the controllers, assuming, for instance, one controller per network. Using this model, it is possible to guarantee that failures in one virtual network do not affect the others.

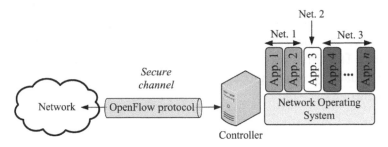

Figure 2.8. *The centralized OpenFlow controller model*

OpenFlow provides a flexible infrastructure based on the idea of distributed forwarding elements already offering basic functions for operating a network, and centralized control planes. Using this infrastructure, it is possible to slice the physical network into multiple virtual networks. In OpenFlow, the instantiation of a network is just the creation of some set of applications in the controller. The new network flows will be created on demand, according to the packets that enter the network. OpenFlow also provides a flexible infrastructure for reallocating network resources. Reallocating a network in OpenFlow means only reprogramming the flow table of each participating node. This is a simple operation for the controller because it knows where the physical devices are and how they are connected.

2.1.3. *Characteristics of Xen and OpenFlow network virtualization technologies*

Neither Xen nor OpenFlow was developed for supporting the Internet pluralist architecture. Nevertheless, up to now, they have been the best commodity alternatives for a virtual network substrate. We evaluate the main characteristics of these technologies, emphasizing their advantages and disadvantages for supporting multiple networks and providing flexibility for innovations.

Xen and OpenFlow have different concepts of virtualization. Xen creates virtual networks by slicing physical network elements into different concurrent virtual routers. Consequently, we consider a virtual network as a set of interconnected virtual routers distributed over the physical infrastructure. On the other hand, OpenFlow creates virtual networks by slicing the network control into many control planes, which sets the forwarding tables of each switch. Hence, using OpenFlow, a virtual network is defined as a set of flows with common characteristics, which are orchestrated by the same set of applications running on the OpenFlow controller. The differences between Xen and OpenFlow virtualization models impact scalability, packet processing, packet forwarding and the use of basic management tools.

2.1.3.1. *Programmability and network processing*

One of the main advantages of the pluralist approach is the innovation support. As a result, the network must be flexible enough to provide end-to-end paths over the available physical infrastructure and to guarantee the administrator the whole control of the network, including the choice of the protocol stack, the forwarding rules, the packet processing, etc.

Because Xen virtualization layer is running directly over the hardware, each virtual router has access to all computer components, such as memory, processor and I/O devices. Therefore, the network administrator is free to choose everything through the virtualization layer. Thus, different operating systems, forwarding tables, forwarding rules and so on can be defined for each virtual network. Furthermore, both data and control planes can be completely virtualized, leading to a powerful and flexible platform for network control and management, as shown in Figure 2.3c). By using Xen, it is possible to add new functionalities, such as hop-by-hop packet processing, which is an important feature toward authentication and access control in the Internet. Adding per-packet signature would solve legacy security problems of the original Internet that could not be changed because of the network "ossification" [CLA 04].

The OpenFlow virtualization model is different from Xen, because the virtual slice is a flow and, the actions are concerned with flows, instead of packets. OpenFlow provides a simple packet forwarding scheme in which the network element performs a flow table lookup to forward an incoming packet. If there is no match, the packet is forwarded to the controller, so that the controller can set a forwarding rule on each node used to forward the corresponding packet. Version 1 of the OpenFlow protocol specifies that the controller can set flow actions, which can result in a header field modification before forwarding the packet. For instance, the forwarding element could change the destination address to forward the packet to a middle box before forwarding it to the next network element. Although flow operations are easily handled by OpenFlow, packet-level features, such as packet signature verification, are not efficient because they must be executed by the controller or by a middle box.

In terms of flexibility, the main disadvantage of OpenFlow is that all virtual networks must forward packets based on the same primitives (flow table lookup, wildcard matching and actions) because there is a single data plane shared by all virtual networks in each network element. On the other hand, Xen provides independent data planes to different virtual networks. To increase the flexibility, OpenFlow provides a fine-grained forwarding table, much more flexible than the table used by the current TCP/IP, adopted by Xen. Currently, Xen provides a forwarding table that is based on IP routing, which means that the forwarding plane is only based on source–destination IP addresses. In contrast, OpenFlow uses an N-dimensional space to define each flow, where N is the number of fields in the header that can be used to specify a flow, as shown in Figure 2.9. Hence, we define a flow based on all the N dimensions or based on a wildcard that narrows the identification of a flow to a subset of relevant header fields [MCK 08]. The consequence of using such a forwarding table is to allow packets to be forwarded based not only on the destination IP, but also on other parameters, such as the application type. Although the same forwarding table can also be implemented in Xen, it is not available yet.

Another key difference between Xen and OpenFlow concerning programmability is the control plane model. In Xen, all virtual nodes have both data and control planes. As a result, the network control is decentralized. In OpenFlow, on the other hand, each node has only the data plane and the control plane is centralized on the controller. Centralizing the control plane simplifies the development of algorithms for network control, compared with a decentralized approach. A centralized control, however, requires an extra server in the network and also introduces a network point of failure.

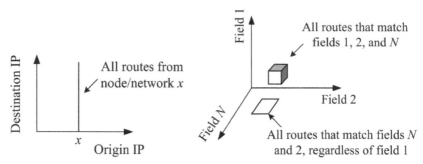

a) Flow definition in the TCP/IP model.   b) Flow definition in the OpenFlow model.

Figure 2.9. *Forwarding table in TCP/IP and in OpenFlow based networks*

2.1.3.2. *Forwarding performance*

One important requirement for the pluralist architecture is being built upon a substrate providing high packet forwarding performance. Packet forwarding depends not only on the hardware used, but also on the logic provided by each technology. In this section, we assume that both Xen and OpenFlow are running on the same hardware to evaluate the losses imposed by each technology on packet forwarding.

As we consider each VM as a virtual router, the performance of Xen acting as a router is a key point. Basically, Xen performance depends on the domain in which the packet forwarding is handled. For each virtual router, packet forwarding can be performed by the operating system running on the domU corresponding to the virtual router or by dom0. In the first case, moving packets between dom0 and domU incurs in control overhead, impacting the overall system performance. In the second case, packets to and from all virtual routers are forwarded by dom0, which simultaneously deals with multiple forwarding tables.

The performance of Xen packet forwarding also depends on the mode employed to move packets between network interfaces: bridge or router modes [EGI 07]. The bridge mode is the default network mode used by Xen, as shown in Figure 2.4. Nevertheless, this mode is not well suited to a router, because it needs more than one physical interface per device. Figure 2.10a) shows an example of the bridge mode with two physical interfaces. We have two bridges on dom0, one per physical interface, connecting the back-end interfaces and the physical interfaces. Packet forwarding, in this case, can be performed at dom0 by using layer-2 or layer-3 protocols. Let p be a packet arriving at the physical interface $ph0$ that must be forwarded to the physical interface $ph1$. First, p is handled by the device driver running on dom0. At this time, p is in $ph0$, which is connected to the bridge $br0$. This bridge demultiplexes the packet p and moves it to the back-end interface $be00$ based on the destination MAC address. After that, p is moved from $be00$ to the front-end interface $fe0$ by using the I/O channel through the hypervisor. The packet p is then forwarded to the front-end interface $fe1$ and, after that, another I/O channel is used to move p to the back-end interface $be01$. This interface is in the same bridge $br1$ of the physical interface $ph1$. Thus, p reaches its outgoing interface. It is worth mentioning that the hypervisor is called twice to forward one packet.

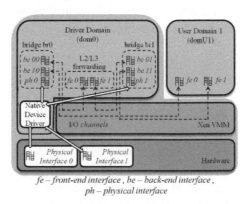

fe – *front-end interface*, be – *back-end interface*,
ph – *physical interface*

a) Bridge mode

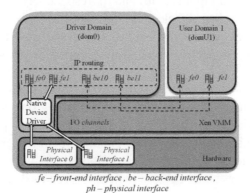

fe – *front-end interface*, be – *back-end interface*,
ph – *physical interface*

b) Router mode

Figure 2.10. *Xen network architectures for packet forwarding*

In the router mode, illustrated in Figure2.10b), the dom0 uses physical interfaces with an IP address associated with each interface. As a result, the router mode does not require bridges connecting each physical interface and I/O channel. Thus, packet forwarding from a physical interface to another interface at dom0 is performed as well as in native Linux. In this case, if dom0 is used as a shared data plane (Figure 2.3b)), there are no calls to the hypervisor. In the router mode, the hypervisor is called only when each virtual router implements its own data plane, as illustrated in Figure 2.3c). In this case, packets are routed to the back-end interface associated with the destination domU and are then moved to the front-end interface by using the I/O channel through the hypervisor. Then, packets are moved to the back-end interface and finally routed to the outgoing physical interface. To allow domUs to send and receive packets, IP addresses are also assigned to back-end interfaces in contrast to the bridge mode.

Using VMs for packet forwarding allows Xen-based virtual networking high programmability. Each network can be configured and can perform processing to decide the actions taken upon each packet received. This characteristic, however, reduces the packet forwarding performance, if compared with a non-virtualized device. To improve packet forwarding, all the required procedures can be executed on dom0, leaving only the control planes to be run on their respective VMs. When packet forwarding is accomplished in dom0, Xen has one control plane for each virtual router and only one data plane shared by all the virtual networks, unlike the conventional model, as illustrated in Figures 2.3b) and c), respectively. The performance of dom0 packet forwarding is close to the performance of non-virtualized devices. The packet processing feature, however, loses flexibility because dom0 does not process packets in a hop-by-hop fashion, as VMs can do.

OpenFlow does not assume virtualized data planes on forwarding elements and, consequently, follows the model of one shared data plane among all networks. Then it is expected that OpenFlow packet forwarding performance is quite similar to native Linux. OpenFlow, however, shows a disadvantage for flows not previously configured. As we commented earlier, when a packet reaches an OpenFlow switch, it is forwarded to the controller if the flow does not match any entry on the forwarding table. The controller has to first configure all the OpenFlow switches used to forward the packet. This mechanism may introduce a considerable delay when forwarding the first packet of each flow. If new flows are continuously instantiated, the OpenFlow performance is affected.

2.1.3.3. Scalability

Scalability is another important issue for a virtualization technology that aims at providing multiple parallel networks. Although native OpenFlow has a better packet forwarding performance than Xen via domUs, Xen is more promising for network scalability, assuming, as the parameter, the number of networking elements. First, OpenFlow assumes that all nodes run a layer-2 protocol at the same level. Therefore, if a node sends an Address Resolution Protocol (ARP) request, all nodes in the network listen to it. OpenFlow, up to now, has not provided support for creating multiple isolated domains in the network. As a result, the current OpenFlow solution is restricted to a local or a metropolitan area network. The Xen architecture is based on the idea that network nodes can run layer-3 protocols. Hence, network nodes operate as virtual routers and establish network domains, which can be hierarchically organized. This structure is compatible with the current network model and is also scalable.

Second, OpenFlow is based on a centralized controller, which configures the network elements. Since the control plane is centralized and the first packet of each flow must be forwarded and processed by the controller, the size of an OpenFlow network is restricted by the processing power and the link capacity of the controller. Up to now, OpenFlow has had no native solution for providing support for different controllers in the same virtual network. Again, Xen model presents a better approach because it is based on a decentralized control plane. Although decentralized algorithms may suffer with slow convergence and are often more difficult in being correctly implemented, they are more suitable when scalability is an issue.

Scalability is also related to the number of virtual routers that can be run over the same physical node. The future Internet requirements are still an open issue and the new architectural design should not restrict the number of virtual networks running over the available physical infrastructure. The Xen approach is less flexible in this sense because the virtual network element is a VM, which demands more hardware resources, such as processing power and memory utilization, than a simple flow in an OpenFlow switch. Context switching and datapath in Xen are much more complex than in OpenFlow. The number of virtual networks in Xen is limited by the hardware of the network element.

Indeed, even if a network element has no traffic at a given moment, it occupies a fixed amount of disk and memory in the physical network element, which may hinder the instantiation of another virtual network element. OpenFlow provides a more flexible infrastructure for the instantiation of a virtual network slice over a physical network element. Since the forwarding network element has only one shared data plane, its resources are not consumed by different virtual operating system or by the maintenance of fixed amounts of memory and disk for specific virtual networks.

As the concept of virtual networks in OpenFlow is given by a set of flows with the same set of characteristics, OpenFlow can support many virtual networks running in parallel. In opposition, Xen is restricted to the number of VMs that can be multiplexed over the network element hardware. It is worth mentioning that Xen scalability can be improved if dom0 is used as a shared data plane.

2.1.3.4. *Basic virtual network management primitives and tools*

Virtual networking management primitives depend on specific tools to create and delete virtual networks, to reallocate virtual networks over the physical infrastructure, to reallocate node resources, and to monitor virtual networks. Xen and OpenFlow present different approaches for carrying out the above-mentioned primitives because they are based on different models and they have different approaches for a virtual network.

Managing virtual networks results in the existence of an entity hierarchically above all the virtual network managers, called, hereinafter, the arbiter. This is an important assumption because if there were no arbiter, each network could try to consume all the available resources, interfering on the operation of the other virtual networks. Indeed, a virtual network management framework also implies the existence of isolation tools, which can be used by the arbiter to guarantee the minimum resources for each network. The presence of an arbiter to decide on the division of the available resources among the coexisting virtual networks is addressed in Feamster *et al.* [FEA 07]. They argue that the Internet service providers (ISP) should be separated from infrastructure providers and, according to them, the infrastructure provider should be responsible for resource sharing among the virtual networks.

The existence of an arbiter raises an issue about security. Since we have an entity that has power over the whole network, the communication between this entity and the virtual/physical nodes must be secure. Moreover, the arbiter cannot be influenced by malicious network nodes that want to divert resources from one network to another. In OpenFlow, the arbiter is naturally defined as the network controller. If the architecture is assumed to use a different controller for different networks, then this arbiter is given by the FlowVisor [SHE 10]. The secure channel between each network node and the controller/FlowVisor is defined in OpenFlow standard. The access control and the trust issues are not treated in the standard and must be implemented. Xen does not provide any kind of arbiter to manage virtual networks because it was not developed for this specific utilization.

Assuming the existence of an arbiter, we can discuss how to implement basic virtual networking operations for management in Xen and OpenFlow. In Xen, we define a network as a set of virtual routers. Hence, creating and deleting a network means, respectively, creating and deleting virtual routers over the same physical infrastructure. Xen provides mechanisms for locally instantiating VMs, assuming that the VM image is already stored in the physical node. Hence, for instantiating a virtual network, the arbiter must first select the virtual infrastructure that will be used, transfer the image of the virtual node to each physical node, and then start the VMs. In OpenFlow, the instantiation of a network does not imply changes on the forwarding node. Indeed, to create a new network, an instantiation of a set of applications in the controller or, in case of using FlowVisor, an instantiation of a new controller is required. The new network flows will be created on demand, according to the packets that reach the network. The selection of the physical resources that will be used by each network is decided on the fly by the controller, or at the moment of network instantiation, in case of using FlowVisor. Neither Xen nor OpenFlow provides algorithms for selecting the best configuration for each coexisting virtual network over the same physical infrastructure.

Another important operation is on-demand virtual network accommodation [PIS 10, CLA 05], which means to reallocate the virtual network if a new network has just been instantiated or if the traffic of a running virtual network has changed. Reallocating a network in Xen can be performed via router migration, instantiation and deletion. If we want to maintain the same virtual topology, which may be an important characteristic to avoid impact on virtual network functions, migration is a mandatory solution because the virtual router is transferred from a physical router to another without changing the virtual topology. A restriction of such a procedure is to ensure that the new physical router has at least the same number of network interfaces as the original router. This can necessitate the construction of tunnels to simulate a one-hop-neighborhood, which does not exist in the physical infrastructure. Moreover, it can incur packet losses, unless some specific mechanisms are in use [WAN 08]. Hence, instantiating and reallocating networks in Xen are challenging operations. OpenFlow, however, provides an easier infrastructure for reallocating network resources because it only requires flow table reconfiguration on each participating virtual network node. This is a simple operation for the controller because it knows where the physical devices are and how they are connected. In OpenFlow, network reallocation is even simpler because of the centralized control plane, which allows a single element to maintain the whole topology and to act upon all nodes. An example of network reallocation in Xen and OpenFlow is shown in Figure 2.11.

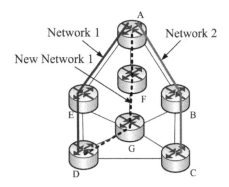

a) Original configuration

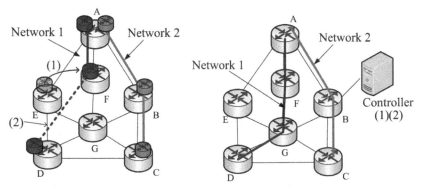

1) Migrate virtual router from E to F   1) Add a flow in nodes D, G, F and A.
2) Create a tunnel from F to D   2) Delete the old flows of nodes A, E, D.

b) Using Xen to migrate Network 1   c) Using OpenFlow to migrate Network 1

Figure 2.11. *Examples of network reallocation using Xen and OpenFlow*

Xen is based on physical node virtualization, leading to resource sharing of machine components such as memory, disk, I/O access and central processing unit (CPU). Indeed, Xen provides tools to manage resource sharing, which can be used to distribute resources, privileging some networks or not. OpenFlow provides less control of the physical node, since the interface between the network node and the controller is rigid. Hence, the control of the physical resources of each network node in OpenFlow is restricted to controlling the frequency of monitoring messages among the node and the controller and the size of the flow table that each network uses. In addition, because the controller/FlowVisor can measure the throughput of each flow, it is also possible to drop flows of a specific network to control network bandwidth. These mechanisms, however, do not provide the same precision as Xen does in controlling each virtual network resource.

2.1.4. Performance evaluation

The performance of Xen and OpenFlow is evaluated in a testbed composed of three machines, as shown in Figure 2.12. The traffic generator (TG) sends packets destined for the traffic receiver (TR) through the traffic forwarder (TF), which simulates a virtual network element. The TF machine is an HP Proliant DL380 G5 server equipped with two Intel Xeon E5440 2.83 GHz processors and 10 GB of random access memory (RAM). Each processor has four cores, and therefore, the TF machine can run eight logical CPUs at the same time. When not mentioned, the TF machine, also called the forwarder, is set up with one logical CPU. The TF machine uses two network interfaces of a PCI-Express x4 Intel Gigabit ET Dual Port Server Adapter. The TG and TR, also called generator and receiver, respectively, are both general-purpose machines equipped with an Intel DP55KG motherboard and an Intel Core I7 860 2.80 GHz processor. The TG and the TR machines are directly connected to the TF via their on-board Intel PRO/1000 PCI-Express network interface.

In the following experiments, packet forwarding performance using native Linux, Xen and OpenFlow is evaluated. Using native Linux, the forwarder runs a Debian Linux kernel version 2.6.26. This kernel is also used in the OpenFlow experiments with an additional kernel module to enable OpenFlow. Using Xen, both dom0 and domUs run a Debian Linux system with a paravirtualized kernel version 2.6.26.

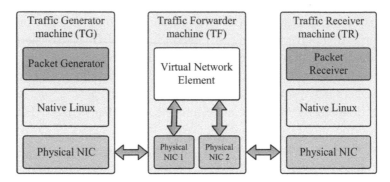

Figure 2.12. *Experimental scenario used in the performance evaluation*

2.1.4.1. *Xen, OpenFlow, and native Linux scenarios*

In the Xen scenario, the three different network configurations are tested. In two of them, Xen runs on the bridge mode; in the first configuration, called XenVM, VMs work as complete virtual routers with both data and control planes on VMs, and in the second configuration, called Xen-Bridge, the VMs contain the control plane whereas the shared data plane is run in dom0. The Xen-Bridge configuration is expected to achieve better packet forwarding performance, even though it reduces the packet processing flexibility if compared with the XenVM configuration. Finally, in the third configuration, Xen runs in the router mode and, in this case, it only evaluates forwarding packets through dom0. This configuration is called Xen-Router. The Xen hypervisor version 3.4.2 is used in all configurations.

In the OpenFlow scenario, the TF acts as an OpenFlow switch. An OpenFlow controller is connected to TF, using a third network interface. TF runs OpenFlow reference system version 0.8.9. The controller is an IBM T42 Laptop that runs a Debian Linux system running NOX version 0.6.0 [GUD 08]. The pyswitch application is used, which is available in NOX to create flow rules in the OpenFlow switch.

In the native Linux scenario, three different packet forwarding configurations are tested. In the first configuration, Native-Router, the forwarder works as a router and runs the standard Linux kernel routing mechanism with static routes. The Native-Bridge configuration uses the Linux kernel bridge, which implements a software-based switch on a PC. Since layer-2 and layer-3 solutions are compared with OpenFlow and Xen, we need to show their performance with both bridge and router modes of native Linux to evaluate the impact of virtualization on packet forwarding. Xen in the bridge mode, however, has a different configuration from the native Linux with bridge. This is because Linux bridge does layer-2 forwarding between two physical interfaces and Xen goes up to layer-3 forwarding. To perform a fair comparison between Xen in the bridge mode and native Linux, a hybrid mode (bridge and router) is provided for native Linux, which is called Native-Hybrid. In this hybrid mode, the forwarder physical network interfaces are connected to different software bridges and a kernel routing mechanism forwards packets between the two bridges. This configuration simulates, in native Linux, what is done in Xen bridge mode as illustrated in Figure 2.10a).

2.1.4.2. *Experimental results*

The packet forwarding rate analysis is conducted with 64-byte (minimum) and 1,512-byte (maximum) frames. Frames with 64 bytes generate high packet rates and force high packet processing in the forwarder, whereas the 1,512-byte frames saturate the 1 Gb/s physical link.

Figure 2.13a) shows the forwarding rate obtained with native Linux, which is an upper bound for Xen and OpenFlow performances. The point-to-point packet rate is also plotted, which is achieved when TG and TR are directly connected. Any rate achieved below the point-to-point packet rate is a consequence of losing packets between TG and TR. Results show that native Linux in the router mode performs as well as the direct point-to-point scenario. This is explained due to the low complexity of the kernel routing mechanism. In the bridge mode, however, native Linux performs worse than the router mode. According to Mateo [MAT 09], this result may be due to the Linux bridge implementation, which is not optimized to support high packet rates. Finally, it is observed that native Linux in the hybrid mode presents the worst forwarding performance. This is expected because of the previously mentioned limitations of the bridge mode and the incremental processing cost required to forward packets from the bridge to the IP layer in the forwarder (TF).

The Xen forwarding rate results are shown in Figure 2.13b). First, a scenario in which dom0 forwards the packets is analyzed. In this scenario, no VM is running, although the same results are expected if VMs are on but not forwarding packets [EGI 07]. In this experiment, the Xen bridge and router modes are analyzed. Xen-Bridge uses the Linux bridge to interconnect VMs, but it also suffers with the same limitations as native Linux in the bridge mode, since the bridge implementation is the same. In addition, Xen-Bridge forwards packets from the bridge to IP layer, as in the hybrid mode, combined with hypervisor calls needed in this mode. As expected, Xen-Bridge performs worse than all native Linux forwarding schemes. On the other hand, Xen-Router performs better than Xen-Bridge, because the Linux bridge is not used and Xen hypervisor is not called when dom0 forwards packets. Nevertheless, Xen-Router is still worse than Native-Router. The forwarding rate rapidly decreases about 1.2 Mp/s load. This behavior is also observed for Xen-Bridge and in the following experiments with VM forwarding. This performance penalty is related to Xen implementation and needs further investigation.

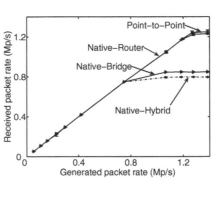

a) Native Linux

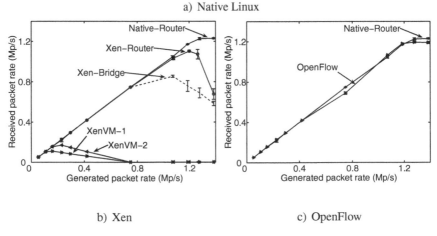

b) Xen   c) OpenFlow

Figure 2.13. *Packet rate for 64-byte frames: native Linux, Xen and OpenFlow*

A scenario in which a VM forwards traffic using Xen bridge mode, which is the default Xen network configuration, is analyzed as well. In XenVM-1, the VM and the dom0 share the same CPU core. This result shows a drop in performance compared with the previous results, in which dom0 was the forwarding element. This poor performance seems to be a consequence of high contention for CPU resources because a single CPU core is shared between both the domains. To eliminate the contention for CPU resources, an experiment with XenVM-2 configuration is accomplished, as seen in Figure 2.13b), in which one exclusive core is allocated to each domain. The performance obtained with the XenVM-2 experiment is better than with XenVM-1, but it is still lower than dom0 results. This can be explained due to the high complexity involving VM packet forwarding. When the traffic is forwarded by the VMs, it must undergo a more complex path before reaching TR. Upon receiving the packet, it is transferred via direct memory access (DMA) to dom0 memory. Domain 0 then demultiplexes the packet to its destination, gets a free memory page associated with the receiving VM, swaps the free page with the page containing the packet and then notifies the VM. To send a packet, a VM must issue a transmission request along with a reference to the memory area where the packet is in the Xen I/O ring. Domain 0 then polls the I/O ring and, when it receives the transmission request, it maps the reference onto the physical page address, and then sends it to the network interface [CHI 07]. This increased complexity is partially responsible for the lower packet rate obtained in the two plots where VMs are used to forward packets.

Figure 2.13c) shows that OpenFlow performs near native Linux in the router mode. In addition, the comparison between OpenFlow and XenVM shows the trade-off between flexibility and performance. Using XenVM, it is possible to achieve more flexibility because data and control planes are under total control of each virtual network administrator. In OpenFlow, however, the flexibility is lower because the data plane is shared among all virtual networks. On the other hand, because of lower processing overhead, OpenFlow performs better than XenVM. Xen performance can be improved if the data plane is moved to dom0, as can be seen in Xen-Router and Xen-Bridge results. In these cases, however, the flexibility of customizing data planes is reduced.

Packet forwarding experiments are also carried out with 1,470-byte data packets, as shown in Figure 2.14. With large packets, all forwarding solutions, but for XenVM-1 and XenVM-2, have the same behavior as the Native-Router scenario. Hence, there is no packet loss in TF and the bottleneck in this case is the 1 Gb/s link. Nevertheless, with XenVM-1, where a VM shares the same core with dom0, the packet rate achieved is lower. In XenVM-2 experiments, in which one exclusive CPU core is allocated to each domain, the behavior is similar to Native-Router. Thus, the performance decrease in the XenVM-1 result is caused by the high contention for CPU resources between domains.

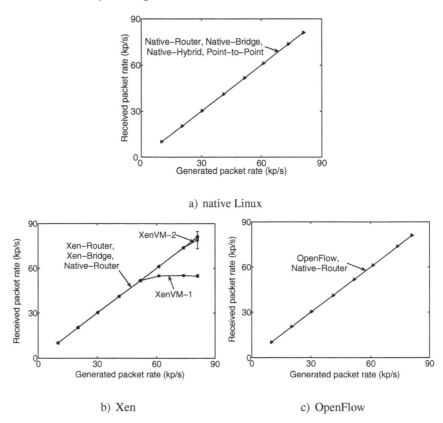

a) native Linux

b) Xen   c) OpenFlow

Figure 2.14. *Packet rate for 1512-byte frames: native Linux, Xen and OpenFlow*

Next, an analysis of the impact of each type of virtual network element on the traffic latency is shown. Background traffic with different rates to be forwarded by the network element is then introduced. For each of those rates, an Internet Control Message Protocol (ICMP) echo request is sent from the generator to the receiver, to evaluate the round trip time (RTT) and the jitter according to the generated background traffic. By measuring the jitter in the ICMP messages, it is investigated if the network element inserts a fixed or a variable delay in the network, which could affect real-time applications.

Figures 2.15a) and b) show the results for the RTT and the jitter, respectively. As the generated traffic increases, the RTT and the jitter measured for ICMP messages increase only for the configuration in which the traffic passes through the VM, called XenVM-1. The difference in the RTT between XenVM-1 and native Linux experiments is up to 1.5 ms in the worst scenario, with background traffic of 500 Mb/s. The RTT and the jitter of OpenFlow have the same order of magnitude as the RTT and jitter of native Linux. Despite the delay difference between XenVM-1 and the other configurations, XenVMs can handle network traffic without a significant impact on the latency. Because the RTT is always smaller than 1.7 ms, even in the worst case, virtual routers running over Xen do not significantly impact real-time applications such as voice over IP (VoIP), which tolerates up to 150 ms delay without disrupting the communication interactivity, even if we consider multiple hops [FAT 05].

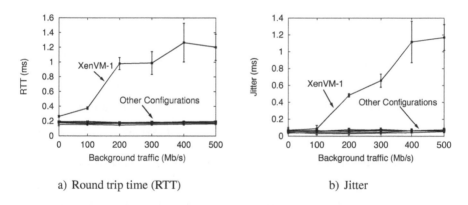

a) Round trip time (RTT)   b) Jitter

Figure 2.15. *Network delay and jitter for 128-byte packets*

Xen and OpenFlow virtualization platform behavior is also analyzed for multiple networks and multiple flows per network. In this scenario, each network is represented as a flow of packets between TG and TR for OpenFlow, and as a VM for Xen. The packet size and the generated packet rate are fixed at 64 bytes and 200 kp/s, respectively. If there is more than one parallel flow, the aggregated generated traffic is still the same. For example, if the experiment is executed with four parallel flows, each flow corresponds to a packet rate of 50 kp/s, generating an aggregated rate of 200 kp/s. Figure 2.16a) shows the aggregated packet rate as a function of the number of virtual networks, with one flow per network. OpenFlow acts as a software switch despite the fact that the first packet of the flow is sent to the OpenFlow controller. The performance obtained is very similar to a software bridge running over native Linux, maintaining the received rate close to the generated 200 kp/s rate. Although Xen dom0 must have its interruptions first handled by the hypervisor, Xen-Bridge performs almost as well as native Linux in the bridge mode. On the other hand, in the case where multiple VMs are simultaneously forwarding traffic (XenVM-1 configuration), the performance degrades as the number of parallel VMs increases. This degradation is mainly because of CPU context switching, which must multiplex the processor among an increasing number of machines, each machine willing to forward its own flow.

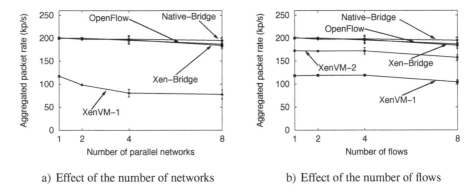

a) Effect of the number of networks   b) Effect of the number of flows

Figure 2.16. *Aggregated packet rate for different number of virtual networks*

Figure 2.16b) shows the aggregated packet rate as a function of the number of flows, considering a single virtual network. As expected, OpenFlow and Xen-Bridge present the same behavior as in Figure 2.16a) because both share the same data plane and, consequently, there is no difference between a virtual network with multiple flows or multiple networks with one flow each. On the other hand, when the traffic is forwarded through the VMs (XenVM-1 configuration), the traffic must undergo a more complex path before reaching TR, as seen in the previous results. To verify if the complex path is the only bottleneck, the test was repeated in a configuration where the VM does not share the same physical core with dom0, referred to as XenVM-2. In this configuration, the performance increases up to 50 kp/s, indicating processing power as an important issue in network virtualization.

To analyze the impact of CPU allocation on VM forwarding, a CPU variation experiment is conducted in which packets are sent from TG to TR at a fixed rate of 200 kp/s through VMs and the number of dedicated CPU cores allocated to dom0 is varied. The 200 kp/s rate is used because, near this rate, it is possible to obtain the best performance in the 1-VM scenario. According to the previous results, the forwarding performance increases when each one, dom0 and VM, has a dedicated CPU core. This test aims at complementing those results by analyzing the forwarding performance when the number of dom0 exclusive CPU cores increases and more VMs forward packets. When more than one VM is used, the aggregated rate of 200 kp/s is equally divided among VMs. Figure 2.17 shows the aggregated received rate in a scenario where each VM has one single core and the number n of CPU cores dedicated to dom0 varies. According to Figure 2.17, the worst performance is obtained when all domains share the same CPU core (i.e. $n = 0$), due to a high contention for CPU resources. As expected, when $n = 1$, the performance increases, because each VM allocates a dedicated CPU core and, consequently, has more time to execute its tasks. In addition, when dom0 receives more than one dedicated CPU core (i.e. $n \geq 2$), the performance is worse when dom0 has a single dedicated CPU core, even when more VMs forward packets. These results show that the network tasks executed by dom0 when each virtual router has two interfaces are single-threaded and these tasks are not performing well in a multi-core environment.

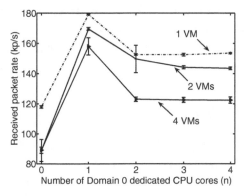

Figure 2.17. *Received packet rate when varying the number of CPUs allocated to Domain 0*

2.2. Xen prototype

A Xen prototype can be developed, such as the one in Grupo de Teleinformática e Automação (GTA) laboratory [PIS 11], to experiment with new interfaces. This section describes the interfaces related to a Xen prototype, developed according to the architecture described in Figure 2.18. The piloting plane requests services to the virtual machine server (VMS) (section 2.2.1). These services can be related either to sense or act on the infrastructure. The VMS performs the action required and sends the answer to the piloting plane. To simplify the implementation of the piloting plane, the interfaces offered by the VMS must be well defined and platform independent.

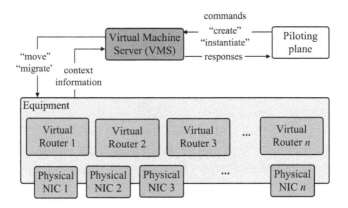

Figure 2.18. *Horizon Xen prototype architecture*

A graphical user interface (GUI) can be developed to help the operation of a Xen prototype (section 2.2.3) to allow the operation of the network by a human. This interface can replace the piloting plane and can have access to all network administrative tasks.

2.2.1. *Virtual machine server (VMS)*

The VMS [ALV 12] provides a set of services for managing virtual networks and virtual routers. This system provides virtual routers on-demand to match specific requirements. Upon receiving requests for new virtual networks, the server creates the correct number of VMs and deploys them in specific nodes of the physical network. Moreover, this server can take part in the network administrative activities.

The server can be implemented using Web services [W3C] and can respond to service requests using the Simple Object Access Protocol (SOAP) [BOX 00]. The Web server can be the Apache Tomcat [APA 10]. This approach simplifies the creation of heterogeneous clients for the VMS and decreases the complexity of adding new features. Each service is a public method modeled as a class called `VirtualMachineServer`.

A piloting plane can act on the network and can autonomously decide to perform some changes, as proposed in the future Internet projects [HOR 10]. For example, we can decrease the load of an overloaded physical machine by migrating one of its machines to another physical host. In this case, the piloting plane sends a command to the VMS using the SOAP protocol to request the migration of a VM. The VMS then uses Libvirt [LIB 10], a virtualized systems management library, to perform the migration operation.

Among the available services in a Xen prototype, there are those that are important to manage the multiple virtual networks. In the following, typical services provided in a Xen-based testbed are summarized for pluralist architectures along with the description of their main goals.

`createVirtualMachine`: this service should be called whenever new VMs must be created on a node of the network.

`createVirtualNetwork`: this service creates a set of VMs on some physical nodes of the network. Moreover, the VM server must map the created virtual network interface onto the indicated physical network interface.

`destroyVirtualMachine`: this service destroys a VM. A destroyed VM cannot be reused in the future.

`registerNodes`: initially, the VM server knows nothing about the physical hosts in the network. This service can be used to register the existing nodes in the network, that is the name, the public key, and the IP addresses of the VMs will be saved at the VM server.

`getPhysicalServerStatus`: this service obtains a list of basic information about the physical server. The current list contains the number of CPUs, the number of cores, the RAM memory size, the amount of free RAM memory, the name of the host and the number and the name of the active virtual domains.

getRegisteredNodes: this service returns a list of the registered nodes on the VM server.

getVirtualMachineStatus: this service returns a list of basic information about a VM. The current list contains the name of the VM, the current RAM memory size, the total RAM memory that can be used, the current number of virtual CPUs (VCPUs), the maximum number of VCPUs that the VM can use, the CPU time used and the current state of the VM.

migrateVirtualMachine: this service migrates a VM from one physical host to another at the same network.

sanityTest: this service is a sanity test for the VM server. The client sends a string to the server and the server sends back the same string.

shutdownVirtualMachine: this service shuts down the VM. In this case, the VM can be used again in the future.

topologyDiscover: this service creates a matrix with the adjacencies on the physical and virtual networks. There is one restriction to this service. A node must be registered on the server, using the service registerNodes, to be part of the physical topology and also to have its VMs on virtual topologies.

getVirtualMachineSchedulerParameters: this service queries the hypervisor for the CPU scheduler, parameters of a VM. For the credit scheduler, the parameters are weight and cap.

setVirtualMachineSchedulerParameters: this service sets the hypervisor for the CPU scheduler parameters of a VM. For the credit scheduler, the parameters are weight and cap.

To use the capabilities offered by the VMS, we have to develop clients for it. A client for the VMS must create a SOAP message with the desired service and its parameters.

2.2.2. *Virtual machine server client*

To simplify client development, a class can be proposed taking into account each new service added to the server. For each new service, a method for the creation of message payloads is added to the client class. In the following, possible messages are listed with their respective payloads implemented in a client class to interact with the VMS. These messages are proposed in Horizon Project [HOR 10].

createVirtualMachinePayload: this method creates a payload to request the creation of a VM, based on the name of the physical machine hosting the new VM, also based on the name of this new VM, on the desired IP address, and on the desired RAM size. The method returns an Extensible Markup Language (XML) message, represented by an object of the class OMElement, with the operation result, which can be either a success or failure.

createVirtualNetworkPayload: this method creates a payload to request the creation of a virtual network, based on a list of the physical machine names that will host the new VMs, on a list with the names of the new VMs, on a list with the desired IP addresses, on a list with the desired RAM memory sizes and on a list of the physical network interfaces that will be mapped onto the new virtual network interfaces created on the VMs. The method returns an XML message, represented by an object of the class OMElement, with the operation result, which can be either a success or failure.

destroyVirtualMachinePayload: this method creates a payload to request the destruction of a VM, based on the name of the physical machine that hosts the VM and on the name of the VM. The method returns an XML message, represented by an object of the class OMElement, with the operation result, which can be either a success or failure.

getPhysicalServerStatusPayload: this method creates a payload to get the physical server status, based on the name of the physical machine. The method returns an XML message, represented by an object of the class OMElement, with the operation result, which can be either a success or failure, the number of CPUs, the number of cores, the total RAM memory size, the amount of free RAM memory, the name of the host, and the number and the name of the active virtual domains.

getRegisteredNodesPayload: this method creates a payload to get the registered nodes, and has no parameters. The method returns an XML message, represented by an object of the class OMElement, with the operation result, which can be either a success or failure, and a list of the registered nodes.

getVirtualMachineStatusPayload: this method creates a payload to get the VM status, based on the name of the physical machine that hosts the

VM, and on the name of the VM. The method returns an XML message, represented by an object of the class OMElement, with the operation result, which can be either a success or failure, the name of the VM, the current RAM memory size, the total RAM memory that can be used, the current number of VCPUs, the maximum number of VCPUs that the VM can use, the CPU time used, and the current state of the VM.

migrateVirtualMachinePayload: this method creates a payload to migrate a VM, based on the name of the source physical machine, the name of the destination physical machine, the name of the VM, and a string indicating if the operation will be a live migration, that is if the migration will occur without the interruption of the programs running on the VM. The method returns an XML message, represented by an object of the class OMElement, with the operation result, which can be either a success or failure.

registerNodesPayload: this method creates a payload to register nodes, based on a list with the physical servers to be registered. The method returns an XML message, represented by an object of the class OMElement, with the operation result, which can be either a success or failure.

sanityTestPayload: this method creates a payload to request a sanity test, based on a string that will be sent to the VM server. The method returns an XML message, represented by an object of the class OMElement, with the string received by the VM server, that is the test is considered well succeeded if the sent and the received strings are the same.

shutdownVirtualMachinePayload: this method creates a payload to shut down a VM, based on the name of the physical machine that hosts the VM, and the name of the VM. The method returns an XML message, represented by an object of the class OMElement, with the operation result, which can be either a success or failure.

topologyDiscoverPayload: this method creates a payload to discover the topology. The service has no parameters. The method returns an XML message, represented by an object of the class OMElement, with the operation result, which can be a either success or failure, and the physical and virtual topologies.

getVirtualMachineSchedulerParametersPayload: this method creates a payload to get VM scheduler parameters, based on the name of the physical machine that hosts the VM and the name of the VM. The method returns an XML message, represented by an object of the class OMElement, with the operation result, which can be either a success or failure, and the values of the credit scheduler parameters (weight and cap).

setVirtualMachineSchedulerParametersPayload: this   method creates a payload to set VM scheduler parameters, based on the name of the physical machine that hosts the VM, the name of the VM to be affected, the new value for the weight parameter, and the new value for the cap parameter. The method returns an XML message, represented by an object of the class OMElement, with the operation result, which can be either a success or failure.

2.2.3. *Graphical user interface*

The GUI of the Xen prototype can access the VMS via command line requests using the client developed as part of the VMS. All the services that the GUI uses are accessed by this Web service communication interface. The command line client receives the services and the parameters needed from the GUI.

2.2.3.1. *Virtual machine server and prototype sensors and actuators*

The VMS (section 2.2.1) offers an integrated interface to control the network Xen machines. Most of the administrative tasks are accomplished using the Libvirt library [LIB 10]. However, there are some tasks that cannot be done only using the Libvirt library, for example topology discovery. For these tasks, a set of additional applications must be developed.

The topology discovery module obtains the virtual and the physical network topologies. In addition, to acquire information regarding CPU, memory and network-related data, another module is needed. Migrating a virtual router without packet losses also requires a module as well as controlling the virtual router throughput. The last one requires a scheduler module to act on the CAP of the VCPUs of the virtual routers. All these additional modules have to obtain data from virtual and physical routers, and also from the controller in order to fully accomplish their objectives. Therefore, to interconnect all of them, a communication module is also needed. The interaction between these modules is shown in Figure 2.19.

Piloting plane interface

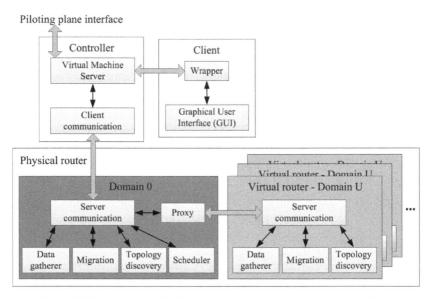

Figure 2.19. *Controller, physical, and virtual routers modules interactions*

Whenever the VMS wants to use one of these applications, it calls the client communication module through command line passing the desired request. If the request is to a physical router, the physical router IP address is provided. If the request is to a virtual router, both the physical router IP address and the virtual router IP address are passed to the client communication module. Next, the client communication module builds an XML message with the request and sends it through a socket to the server communication module running inside the physical router. If the request regards the physical router, the server communication module creates an instance of the correct application to handle the request, calls the application with the provided parameters and returns to the client communication module an XML containing the application response. If the request regards a virtual router, then the physical server acts as an intermediate between the controller and the virtual router. In this case, the server communication module de-encapsulates the message sent by the client communication module and handles it to the proxy module, which is responsible for sending the message to the server communication module of the right virtual router. When the message arrives at the server communication module in the virtual router, an instance of the appropriate application is created, it handles the request and the application response is sent back to the controller through the virtual router server communication module, the proxy module and the physical router server communication module.

2.3. OpenFlow prototype

An OpenFlow prototype was deployed in GTA laboratory [MAT 11], to experiment new interfaces. This section describes the components related to an OpenFlow prototype.

Following the same idea of the Xen prototype, the OpenFlow prototype can also be based on Web services. The communication between the core prototype and the external applications uses the HTTP (HyperText Transfer Ptotocol) to exchange XML messages. To measure the OpenFlow network performance, sensors can be used. These sensors are basically counters installed on switches, accessible via the OpenFlow protocol, or information on the OpenFlow table, such as the number of flow entries and other statistics. NOX applications collect sensor information and make them available as a Web service.

Figure 2.20 shows the OpenFlow prototype architecture. The NOX controller is the base of OpenFlow applications. The NOX controller provides the OpenFlow protocol and the secure channel implementation for the applications running atop.

2.3.1. *Applications*

A list of applications running on a NOX controller along with their main goals is described below.

Stats application: this application collects statistics about switches and converts them into an XML message.

Discovery application: this application discovers the network topology and describes it as an XML message.

SpanningTree application: this application implements a spanning tree algorithm that avoids the occurrence of network loops. The topology of the defined spanning tree is available as an XML message.

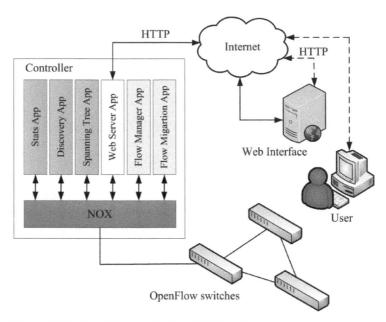

Figure 2.20. *OpenFlow applications, NOX and interactions among agents*

FlowManager application: this application implements flow changes by adding, modifying and deleting flows.

FlowMigration application: this application implements flow changes by migrating a flow from one path to another.

WebServer application: this application provides the integration among features of NOX controller applications. The WebServer App implements the HTTP protocol to provide an interface between NOX controller applications and external applications. Consequently, the NOX controller application handles the HTTP requests, converts them in an application method call and executes the method. The prototype used for the experiments implements a client for the WebServer App. The client is another Web server that enables administrators to control the OpenFlow network. This client takes part in a Web application that provides a GUI, described later in section 2.3.3.

2.3.2. *OpenFlow Web server*

The WebServer App [MAT 12] is a NOX application that is responsible for providing a Web interface for other NOX applications. The WebServer

App implements the concept of Web service, in which a functionality of other applications can be accessed by an HTTP request, and returns XML messages. The implementation of this application is based on the NOX Web server default application, which can be set to run on 8080 port, listening HTTP requests. All HTTP requests are handled by the OpenFlow resource, defined on mywebserver application. For each Web service provided by the OpenFlow resource, there is a method defined in the MyWebServerResource class. The concepts concerning each WebServer App component are described below.

The NOX default WebServer App implements a framework for deploying Websites as a NOX application. This feature is used to implement the Web services as a special kind of Websites. The mywebserver class is a NOX application for class implementation. It starts all applications that must run simultaneously with the WebServer App. It also starts the NOX default WebServer App and defines its default resource as an object of the class MyWebServerResource. The MyWebServerResource class defines a resource, which is a kind of Website that is implemented by the NOX default WebServer App . This class also implements the mapping of URL requests to function calls, providing an interface between the user and the OpenFlow network. There are some services already implemented on MyWebServerResource. Each service can be accessed by an HTTP request using a specific URL. The services are described as follows.

getStats: this service does not take any parameters. It calls the Stats App and returns the statistics and the information about the OpenFlow switch network in an XML message.

getTopology: this service does not take any parameters. It calls the Discovery App and returns the topology of the network in an XML message. This service returns a list of all network links.

getNeighbor: this service does not take any parameters. It calls the Discovery App and returns the topology of the network in an XML message. This service returns the list of all node neighbors for each node in the network.

getSpanningTree: this service does not take any parameters. It calls the Discovery App and returns the spanning tree of the network in an XML message. This service returns the list of the node neighbors, which are linked to the node by a spanning tree link, for each node in the network.

addFlow: this service takes flow characteristics as parameters, like the flow match, idle timeout, hard timeout, priority and action. This service adds a new flow calling the FlowManager App, which performs the required action over the network.

delFlow: this service takes as parameters flow characteristics as parameters, like the flow match, idle timeout, hard timeout and priority. This service deletes a flow calling the FlowManager App, which performs the required action over the network.

migrateFlow: this service takes flow characteristics as parameters, like the flow match, idle timeout, hard timeout, priority, action and also the list of switches on which the flow must be set. This service migrates a flow calling the FlowMigration App, which performs the required action over the network.

2.3.3. *Graphical user interface*

OpenFlow switches forward network traffic according to a flow table containing the active flows. This table contains flow characteristics and rules to be applied, such as determining the queue and the output port. This table can be configured locally or by a network controller. A user-friendly interface can be implemented to allow users to modify flow tables in order to facilitate configuration and network management. This user interface was developed based on a Web application in which, using a Web browser, the user can access the interface and can run commands and queries to manage the network.

The application that provides the GUI is divided into three layers. The first layer is the *Data Layer*, where user commands are executed and where the collection of data is performed as responses to queries. The second layer is the *Data Processing Layer* that processes all received information before sending it to other layers. The third layer is the *Presentation Layer* that organizes and shows the data to the user. The isolation provided by layer structure allows the modification of a specific layer without modifying the others. Figure 2.21 shows the layers of the application and the protocol used to exchange messages.

Data layer: the Data Layer is composed of a NOX controller and its applications. The application WebServer App is responsible for providing the communication interface between the Data Layer and the Data Processing Layer. This application communicates with Data Processing Layer over the HTTP protocol.

Data processing layer: the Data Processing Layer is composed of a Web server, which processes requests and commands from users. Designing our own server, instead of using an existing server such as Apache, may be useful to offer administrators full control of the provided services and, as a result, it facilitates the system development.

Data presentation layer: the Presentation Layer is composed of mark and script files that are interpreted by the Web browser, which can be Hyper Text Markup Language (HTML), JavaScript (JS), Cascading Style Sheets (CSS), XML and Scalable Vector Graphics (SVG) files. HTML files have the Web page description that a Web browser interprets in order to show this Web page.

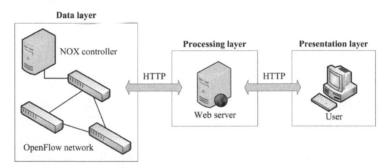

Figure 2.21. *Application layers*

CSS files have the style markup to improve the presentation provided by HTML. A CSS file is interpreted by the Web browser and the style is applied to HTML markup. The JS files have the script functions to make Web pages dynamic and interactive. The XML files have information that Web browser or JS functions use in order to show or exchange data. The messages that are provided as user command responses are XML messages.

SVG files have an XML description of network topology to provide graphic visualization. A combination between SVG and JS allows animations and interactions with the image generated by SVG.

In this layer, there are resources that allow users to acquire network information and execute commands.

2.4. Summary

In this chapter, performance measurements are conducted using two virtualization platforms, OpenFlow and Xen. The results presented show that Xen enables a highly flexible environment, with different protocol stacks running in parallel using customized network-data forwarding structures and lookup algorithms. This flexibility has a high performance cost limiting the VM packet forwarding capacity to less than 200 kp/s. On the other hand, OpenFlow shows a packet forwarding performance similar to a native Linux environment.

A prototype for each virtualization platform was introduced as well as the interfaces that can be used on each platform. For the Xen platform, a VMS was developed using the Web service concept. Using the VMS, physical and virtual hosts of the network can be controlled. To simplify the network administration by a human agent, a GUI can be developed. This interface can be used to show the topology of the network and act on its elements.

In the OpenFlow prototype, similar tools can also be developed. The WebServer App offers a Web interface to agents interested in administrating the OpenFlow network. A GUI can also be developed. This interface can be accessed using a Web browser to view information related to the OpenFlow network. More services can be added to both virtualization platforms, Xen and OpenFlow. Moreover, other virtualization platforms can be proposed and similar services can be added and compared with the existing platforms.

2.5. Bibliography

[ALV 12] ALVES R.S., CAMPISTA M.E.M., COSTA L.H.M.K., et al., "Towards a pluralist internet using a virtual machine server for network customization", in *Asian Internet Engineering Conference (AINTEC'2012)*, Bangkok, Thailand, pp. 9–16, November 2012.

[AND 05] ANDERSON T., PETERSON L., SHENKER S., et al., "Overcoming the internet impasse through virtualization", *IEEE Computer*, vol. 38, pp. 34–41, April 2005.

[APA 10] "APACHE TOMCAT", March 2010. Available at http://tomcat.apache.org/ (accessed in May 2013).

[BOX 00] BOX D., EHNEBUSKE D., KAKIVAYA G., *et al.*, *Simple Object Access Protocol (SOAP) 1.1*, RFC 3288, March 2000.

[CHI 07] CHISNALL D., *The Definitive Guide to the Xen Hypervisor*, Prentice Hall, 2007.

[CLA 04] CLARK D., BRADEN R., SOLLINS K., *et al.*, New arch: future generation internet architecture, Technical report, USC Information Sciences Institute Computer Networks Division, MIT Laboratory for Computer Science and International Computer Science Institute (ICSI), August 2004.

[CLA 05] CLARK C., FRASER K., HAND S., *et al.*, "Live migration of virtual machines", *Usenix Symposium on Networked Systems Design and Implementation (NSDI)*, Usenix, pp. 273–286, May 2005.

[EGI 07] EGI N., GREENHALGH A., HANDLEY M., *et al.*, "Evaluating Xen for router virtualization", *International Conference on Computer Communications and Networks (ICCCN)*, Honolulu, USA, pp. 1256–1261, 13–16 August 2007.

[FAT 05] FATHI H., PRASAD R., CHAKRABORTY S., "Mobility management for VoIP in 3G systems: evaluation of low-latency handoff schemes", *IEEE Wireless Communications*, vol. 12, pp. 96–104, April 2005.

[FEA 07] FEAMSTER N., GAO L., REXFORD J., "How to lease the internet in your spare time", *ACM SIGCOMM Computer Communication Review*, vol. 37, pp. 61–64, January 2007.

[FER 11] FERNANDES N.C., MOREIRA M.D.D., MORAES I.M., *et al.*, "Virtual networks: isolation, performance, and trends", *ACM SIGCOMM Computer Communication Review*, vol. 66, pp. 339–355, February 2011.

[GUD 08] GUDE N., KOPONEN T., PETTIT J., *et al.*, "NOX: towards an operating system for networks", *ACM SIGCOMM Computer Communication Review*, vol. 38, pp. 105–110, July 2008.

[HOR 10] "Horizon Project: a new horizon to the Internet", March 2010. Available at http://www.gta.ufrj.br/horizon.

[LIB 10] "Libvirt: the virtualization api", June 2010. Available at http://libvirt.org/.

[MAC 11] MACEDO D.F., MOVAHEDI Z., RUBIO-LOYOLA J., *et al.*, "The AutoI approach for the orchestration of autonomic networks", *Annals of Telecommunications*, vol. 66, pp. 243–255, April 2011.

[MAT 09] MATEO M.P., OpenFlow switching performance, Master's thesis, Politecnico Di Torino, Torino, Italy, July 2009.

[MAT 11] MATTOS D.F., FERNANDES N.C., DA COSTA V.T., *et al.*, "OMNI: Open-flow MaNagement Infrastructure", *IFIP International Conference on the Network of the Future (NoF)*, Paris, France, pp. 1–5, November 2011.

[MAT 12] MATTOS D.M.F., FERNANDES N.C., DA COSTA V.T, *et al.*, "OMNI: OpenFlow MaNagement Infrastructure", in *2nd IFIP International Conference Network of the Future (NoF'2011)*, Paris, France, November 2011.

[MCK 08] MCKEOWN N., ANDERSON T., BALAKRISHNAN H., *et al.*, "OpenFlow: enabling innovation in campus networks", *ACM SIGCOMM Computer Communication Review*, vol. 38, pp. 69–74, April 2008.

[MEN 06] MENON A., COX A.L., ZWAENEPOEL W., "Optimizing network virtualization in Xen", *USENIX Annual Technical Conference*, Boston, USA, pp. 15–28, May 2006.

[PFA 09] PFAFF B., HELLER B., TALAYCO D., *et al.*, OpenFlow switch specification version 1.0.0 (wire protocol 0x01), Technical report, Stanford University, December 2009.

[PIS 10] PISA P.S., FERNANDES N.C., CARVALHO H.E.T., *et al.*, "Open-Flow and Xen-based virtual network migration", *IFIP International Conference on the Network of the Future Conference (NoF)*, Brisbane, Australia, pp. 170–181, September 2010.

[PIS 11] PISA P.S., COUTO R.S., CARVALHO H.E.T., *et al.*, "VNEXT: virtual NEtwork management for Xen-based Testbeds", *IFIP International Conference Network of the Future (NoF)*, Paris, France, pp. 1–5, November 2011.

[SHE 10] SHERWOOD R., CHAN M., COVINGTON A., *et al.*, "Carving research slices out of your production networks with OpenFlow", *ACM SIGCOMM Computer Communication Review*, vol. 40, pp. 129–130, January 2010.

[VER 07] VERDI F.L., MAGALHÃES M.F., MADEIRA E., *et al.*, "Using virtualization to provide interdomain QoS-enabled routing", *Journal of Networks*, vol. 2, pp. 23–32, April 2007.

[WAN 08] WANG Y., KELLER E., BISKEBORN B., *et al.*, "Virtual routers on the move: live router migration as a networkmanagement primitive", *ACM Conference of the Special Interest Group on Data Communication (SIGCOMM 08)*, Seattle, Washington, USA, pp. 231–242, August 2008.

[W3C] W3C *Web Services Activity*, March 2010.

Chapter 3

Performance Improvement and Control of Virtual Network Elements

Chapter 2 has defined the five primitives (instantiate, delete, migrate, monitor and set) that the network virtualization infrastructure must provide to allow the piloting plane to control and manage virtual network elements [SEN 10]. Figure 3.1 shows the relationship between the piloting plane and a general virtualized network element.

primitives

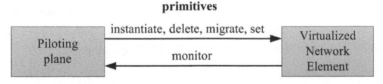

Figure 3.1. *The control primitives and the relationship with the piloting plane and the virtualized network element*

Basically, the piloting plane runs intelligent algorithms [SOA 10, FRE 10] to automatically instantiate/delete virtual networks and also to migrate network elements and set its resource allocation parameters [VAN 10, ALK 10]. Hence, the piloting plane needs to acquire information and uses the monitor primitive. This primitive executes calls to the monitoring

Chapter written by Igor M. MORAES.

tools required to measure variables of interest, such as available bandwidth, processor and memory usage, link and end-to-end delay. After monitoring, the piloting plane is able to act on the network by using the four primitives. The parameters available to monitor and to set include low-level and hardware-specific parameters, such as the number of virtual processors given to a virtual router and priority of processor usage in a contention scenario. Thus, the piloting plane dynamically adapts the resources allocated to each virtual network according to the current network state, number of users, priority of each virtual network, Service Level Agreements (SLAs), etc.

We consider that a virtual network element has two main planes: the virtualization and the piloting planes. The virtualization plane provides the substrate for running logical network elements over a single shared physical network. The piloting plane provides the intelligence for optimizing network performance based on SLAs for each virtual network. Figure 3.2 shows the architecture of a general virtual network element such as the architecture presented in Chapter 2. The core of the node is the virtualization system, which may be implemented by different virtualization tools such as Xen or OpenFlow. Sensors and actuators also compose the virtualization system. Sensors collect information enabling the description of the context they belong to and actuators are software components that apply actions required by the piloting plane on the network element. In Figure 3.2, the controller represents the piloting plane, which receives and aggregates information sent by nodes and, then, sends commands to set parameters and to perform control actions on nodes. All nodes must provide an interface with the piloting plane and should be able to exchange Extensible Markup Language (XML) control messages with it.

This chapter focuses on the specific modifications required by the virtualization system to implement the five primitives used by the piloting plane and also to improve the performance of virtualized network elements. Each virtualization tool introduces a different set of sensors and also requires different mechanisms to manage the network and to implement the primitives of the piloting plane. For example, Xen provides built-in actuators to create and destroy virtual machines (VMs), but the native VM migration mechanism is not well adapted to virtual router applications because it does not avoid packet losses. Therefore, an efficient migration mechanism must be provided. Similar to Xen, OpenFlow provides actuators to create and destroy flows, but it also lacks a native flow migration mechanism. Thus, implementation improvements are detailed for two specific prototypes based on Xen [PIS 11] and OpenFlow [MAT 11] virtualization tools[1].

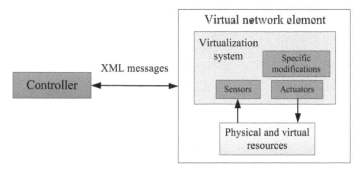

Figure 3.2. *Architecture of a general virtual network element: the controller receives data acquired by sensors and sends commands to manage physical and virtual resources of network elements*

3.1. Xen-based prototype

The Xen-based network prototype is composed of several modules that run on different types of nodes. Basically, each node plays a different role: controller, router or client. The controller is a special node that receives and consolidates data received from all network nodes. The controller node acquires data and sends commands to physical and virtual routers. Router nodes are the network substrates. One physical router runs one or more virtual routers. Both physical and virtual routers run several modules that monitor and promptly act upon the reception of commands. Finally, the client node allows users to interact with the controller by using a Graphical User Interface (GUI). This GUI presents prototype information for user monitoring and also a simple control interface. Figure 3.3 details the main modules of the Xen-based prototype and its interfaces.

The controller has two modules: the Virtual Machine Server (VMS) [ALV 12] and client communication. The VMS is the core of the controller. It has a Simple Object Access Protocol (SOAP) interface for interacting with both the piloting plane and the client node. VMS consolidates all prototype information and executes all control and maintenance algorithms. The

1 Both prototypes are developed by the Horizon Project Team (http://www.gta.ufrj.br/horizon).

detailed description of VMS is presented in section 2.2.1. The client communication module is used by VMS to communicate with other router-node modules that are running on routers.

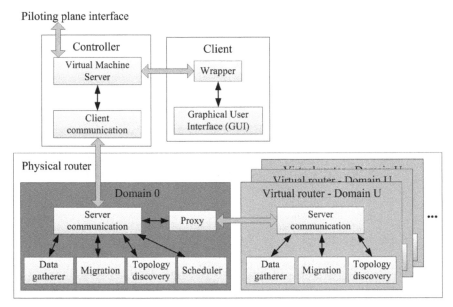

Figure 3.3. *The Xen-based prototype architecture: main modules and relationships*

Physical routers provide the substrate used by virtual networks. Each physical router executes the Server Communication module that receives requests from the controller and forwards these requests to specific modules. If a request is addressed to a given virtual router, it is forwarded to the specific virtual router by using the Proxy module. Otherwise, it is forwarded to one of the other modules running on the physical router. For example, if the received request regards a monitoring task, it is forwarded to the Data Gatherer module to acquire measures from the physical and virtual routers. If the received request corresponds to an action related to resource sharing among virtual routers, the Scheduler module handles this request. The Topology Discovery module handles requests related to the physical or virtual networks topology discovery. Migration requests are handled by our Migration module, which provides virtual router migration without packet losses, as detailed in section 3.1.1.

The client node allows users to monitor and control the network prototype. This node has a GUI that shows both physical and virtual network topologies and additionally allows fine-grain monitoring by exhibiting detailed information about a selected physical or virtual router. The client node also allows users to send commands to the prototype. For example, users can execute the migration of a virtual router from one physical router to another physical router through a mouse click. The GUI running on the client interacts with the controller by using the Wrapper module. This module converts commands from GUI into SOAP and then calls the controller. The Wrapper module also converts SOAP responses from the controller into response patterns expected to GUI [KAR 10].

The following sections describe the improvements on Xen implementation that are adopted in the network prototype. Section 3.1.1 describes the native Xen migration mechanism and introduces our mechanism that allows virtual router migration with no packet losses. Section 3.1.2 presents our tools for network monitoring. Section 3.1.3 describes the module responsible for discovering the physical and virtual network topologies. Finally, section 3.1.4 presents the new virtualization hardware support technologies that are currently available to improve Input/Output (I/O) virtualization performance and its impact on virtualized routers.

3.1.1. *Xen migration*

The migration primitive is used to redistribute physical resources among virtual routers. The idea is to move virtual networks/routers among the available physical resources without disrupting the packet forwarding and network control services [WAN 08]. A special case of the migration is the live migration that allows the dynamic reconfiguration of the network topologies without shutting down the running routers. Thus, live migration allows dynamic planning of physical resources and traffic management on demand.

We analyze the standard Xen migration scheme and we also propose a new migration model that fixes the issues that were found on the standard mechanism. The migration is an important control primitive because it enables the piloting plane to dynamically rearrange the logical network topology, without disrupting running services and with no packet loss. After the formal definition of these technologies, we present a detailed documentation of the prototype with its overview and functionalities.

Xen has a native migration mechanism developed to move VMs [CLA 05]. This mechanism is based on two assumptions: (1) the migration occurs inside a local network and (2) the VM hard disk is shared over the network[2]. Thus, the VM migration is composed of two main procedures: (1) the copy of VM memory to the new physical location and (2) the reconfiguration of the network links without breaking connections.

There are several alternatives to perform the memory copy. The simplest alternative is to suspend the VM, transfer all memory pages to the new physical node and then resume the VM. Although simple, this alternative suffers high downtime, which is the time that a VM is unavailable during the migration. To reduce the downtime, the suspend-transfer-resume procedure evolved to pre-copy migration that has two phases: (1) iterative pre-copy and (2) stop-and-copy. During the first phase, all memory pages are transferred to the new physical machine except those called "hot pages", that is the most frequently modified pages. Thus, the downtime is reduced because a few number of pages, only the hot pages and not all memory pages, are transferred while the VM is down. The iterative pre-copy phase works as follows. In the first round, all memory pages are transferred from the source to the destination machine with a minimum transfer rate specified by the network administrator. Then, in the following rounds, only memory pages that were dirtied during the previous round by the operating system will be transferred. The transfer rate is updated at each round according to an adaptive mechanism based on the "dirtying rate". In each round, the dirtying rate is calculated as the ratio of number of dirtied pages on the last round and the duration of the last round. The maximum rate of the next round is then obtained by adding a constant increment of 50 Mb/s to the calculated dirtying rate. The pre-copy ends if the maximum rate of the round is equal to the maximum rate specified by the administrator or less than 256 kB of dirtied pages remains to be transferred. After that, the stop-and-copy phase starts by suspending the VM and sending the hot pages with the maximum transfer rate to the destination node. Then, the process ends when the destination node confirms the reception of the whole memory to the old physical node.

---

2 This shared disk assumption can be relaxed if both source and destination nodes implement the same small set of applications [WAN 08]. Then, the destination node is able to load these applications onto the file system of the new VM. Hence, only VM memory and configuration files must be migrated.

It is clear that the Xen native VM-migration mechanism works well for server consolidation applications, but it is not efficient for virtual router migration. By using the pre-copy mechanism, the downtime is in the order of hundreds of milliseconds and during this time packets are lost. Depending on the data transmission rate, virtual routers experience high number of packet losses during downtime. It is also important to minimize the total migration time to guarantee that we can quickly free the resources of the source physical machine. Another problem of the Xen native mechanism for migrating virtual routers is that it assumes that the migration is always from one physical router to another physical router in the same local area network (LAN). In the Internet, we cannot assume that physical nodes always belong to the same local network.

To migrate virtual routers with no packet loss, an alternative is to implement the plane separation technique in Xen [PIS 10]. As explained in Chapter 2, a network element has two planes: the control and the data planes. The control plane runs all the control algorithms and builds the routing table, while the data plane forwards packets. With Xen, both planes are implemented within VMs (or by user domains – Domain U). In our prototype, the control plane remains in VM, while the data plane is implemented in Domain 0. Each virtual router has its own forwarding table in Domain 0 and each table is a copy of the original forwarding table created by the routing software running in the VM. When Domain 0 receives a control message, it checks which network the message belongs to and forwards the message to the corresponding VM. When Domain 0 receives a data message, it is forwarded by this domain based on a forwarding table that corresponds to that virtual network.

The migration mechanism with plane separation works as follows. First, the procedure starts in the same manner as the native migration mechanism: the iterative pre-copy phase is executed, the VM is paused and the remaining dirty memory pages are transferred to the new physical router. During this time, the data path continues to work at Domain 0 in the source physical router. Therefore, there are neither interruptions nor packet losses until this phase. Different from the native mechanism, the zero packet loss modified mechanism runs a daemon in Domain 0 to buffer the control packets for the VM that is being migrated. When the whole memory is copied, the VM is resumed on the new physical machine and the network connections are created in this new Domain 0 using a dynamic interface binding module, which maps the virtual network interfaces onto the physical network interfaces of the new physical router. Next, a tunnel from the source to the new physical machine is created in order to transfer the control packets that were buffered in the source Domain 0 and also the new control packets. Finally, the Address Resolution Protocol (ARP) reply is broadcast to update the links and the data path in the old physical machine is removed.

The migration mechanism with plane separation guarantees no packet loss in data plane during the VM migration. There is also no control packet loss. The mechanism inserts only a delay in the control packet delivery. The modified mechanism, however, is based on Xen default migration, i.e. it also requires that routers be within the same LAN. In addition, the mapping of a virtual link over multiple physical links is still an open issue that depends on solutions such as Internet Protocol (IP) tunnels or instantiating new virtual routers on the network. For instance, in Figure 3.4, we migrate Virtual Node B from Physical Node 2 to Physical Node 6. Physical Node 6, however, is not a one-hop neighbor of Physical Node 1. Consequently, to complete the link migration, we need to create a tunnel from Physical Node 6 to Physical Node 1 to simulate a one-hop neighborhood. The other solution is to instantiate a new virtual router to replace the tunnel. This solution, however, modifies the virtual topology and impacts the routing protocol operation.

The modifications on native Xen to implement plane separation are the following. First, Domain 0 should know all VMs that are running in its same physical machine in order to send control messages to the correct VM. Furthermore, control and data planes should interact to maintain the forwarding table consistent. For that, we create two components: the Hello component and the Route Change component. The Hello component sends hostname and network interfaces information of each VM to Domain 0. Then, Domain 0 creates a route table for this virtual router and the rules for using this table. All the packets of this router are thus forwarded by Domain 0 using the virtual router specific routing table. After that, control plane monitoring starts. When a route is changed, this change are also executed in Domain 0. The Route Change component monitors route modifications. There is also another component that is used during the migration process, the Migration Advise component. This component runs on both source and destination Domains 0 and reports the success or the failure of the migration process. The router keeps forwarding packets using data plane in this Domain 0 until the network manager actually decides to migrate it to another physical machine. The standard Xen migration procedure is the prototype first step of the zero-loss prototype procedure. By using this mechanism, we migrate the control plane, but we keep the data plane in its current physical machine, also called source Domain 0. After that, the forwarding environment, the route table and the rules, of this virtual router is created on the destination physical machine, also called destination Domain 0. The new route table is populated with the control plane routes in the migrated VM. At this moment, we have the source Domain 0 forwarding the packets and the destination Domain 0 ready to forward. We start migration of links. To migrate the links, we use the ARP mechanism, which forces our prototype to map one logical hop onto two or more physical hops. Therefore, our migration takes place between machines that have connectivity with all the logical neighbors. The ARP reply message is sent for notifying the Medium Access Control (MAC) address of the interface with the given IP address. Hence, the destination physical machine sends the ARP reply message for each interface used by this virtual router, notifying the neighbors that the router is currently in another location. Thereafter, all the links are migrated and the data plane that is running over the source Domain 0 is dropped.

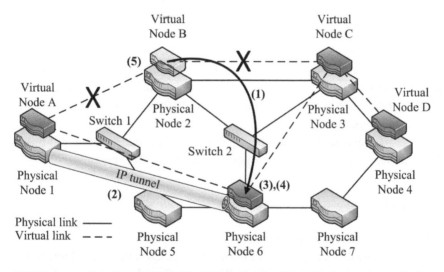

(1) Migrate control plane from Physical Node 2 to 6 maintaining data and control planes
(2) Do interface dynamic binding and create tunnel between Physical Nodes 1 and 6
(3) Create a new forwarding table in Domain 0 of Physical Node 6
(4) Reconfigure links with ARP reply
(5) Delete data plane on Physical Node 2

Figure 3.4. *Example of a Xen-based router migration when a virtual link is mapped into a multihop path in the physical network*

3.1.2. *Xen statistics*

Statistics are required by the piloting plane to manage the virtual routers and networks. Xen provides a few tools to acquire limited information about VMs. Thus, we have developed measurement tools to obtain more general information about virtual and physical routers, virtual networks and networks' substrate. This set of tools composes the Data Gatherer module. Basically, this module retrieves information about resource allocation and resource usage of Domains 0s and Domains Us. With this information, the prototype can be aware of the resource allocation status of the entire network.

The Data Gatherer module is composed of several components that are specialized in gathering information from different measurement tools, as shown Figure 3.5. The components are: Xentop Gatherer, Ifconfig Gatherer, Memory Gatherer, and Latency Gatherer. Each component is responsible for acquiring a set of measurements. For example, the Xentop Gatherer component gathers information acquired by the xentop tool, which includes domain name, central processing unit (CPU) and memory usage, number of virtual CPUs per VM, etc. The Xentop Gatherer module executes only in Domain 0 and provides non-invasive information from all the domains running inside the physical machine. The Data Gatherer module has also two special components: the Data Gatherer Handler, which is responsible for allowing the communication with the other modules of the Xen prototype, and the Data Gatherer main component, which is responsible for calling specific measurement modules and consolidating the output to fulfill the requests. The services provided by the data Gatherer can be accessed through XML message requests and the response is also an XML message.

3.1.3. *Xen topology*

The Topology module discovers both physical and virtual network topologies. This module probes all neighbors[3] of each network element using the Nmap Security Scanner [WOL 02]. The Topology module has three components: Scanning Neighbors, Topology Consolidate and Node Consolidate.

---

3 The term "neighbor" means those nodes that have direct interconnection through one of the network interfaces of the network element.

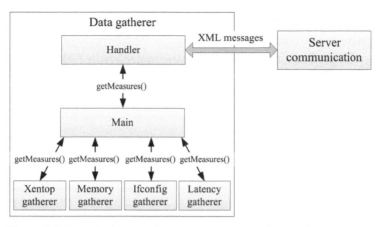

Figure 3.5. *Data Gatherer architecture: the main module sends requests to specific monitoring modules and then forwards acquired data to the handler that processes these data*

The Scanning component is responsible for discovering all the neighbors of a given network element. Both physical and virtual network elements run this component. The neighbor discovery uses the Nmap tool, which probes all the IP range of each network interface of the network element. We add to the neighborhood list all IP addresses that respond to the probes. After that, we have all the neighbors of the network element, which are connected by all network interfaces. The neighborhood information has the neighbor's IP and MAC addresses and the latency of the link. After discovering the neighbors of all network interfaces, the component creates the neighborhood list and transfers it to the Node Consolidate component through XML messages, as shown Figure 3.6.

The Node Consolidate component runs on each physical element of the network. This component aims at obtaining the information about the neighbors of this element in both physical and virtual networks. The Topology Consolidate component calls the Node Consolidate component, which has three tasks. The first task is to discover the physical neighbors of the network element. The second task is to discover all the neighbors of each virtual network element running over this physical network element. Both discovery procedures use the Scanning component. The first two tasks run at the same time because there is no dependency between them. The last task of the Node Consolidate component is to consolidate information. The consolidation consists of the neighborhood information of the physical network element and a list of virtual network elements with its neighbors. The module returns the consolidated information to the controller, which runs the Topology Consolidate component using the messages in XML format.

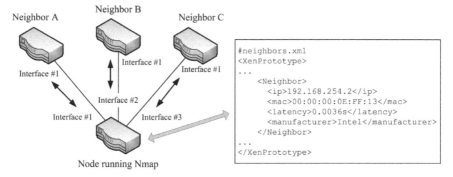

Figure 3.6. *The Scanning Neighbors component: nmap tool is used to probe all interfaces of the network element and thus the nodes that respond these probes are added to the neighborhood list*

The VMS has a list of the registered nodes of the physical network, but there is no information about the interconnections among these nodes. The Topology Module provides the interconnection information. This module asks every physical node in the registered node list about the interconnection information of the physical and virtual elements. After receiving all the interconnection information, the Topology Consolidate component calculates the topology of each network. We model our network topology as a graph. Therefore, the Topology Consolidate component returns a graph representation to the VMS. The graph is represented by the adjacency matrix of each network. Thus, the Topology module allows the VMS to provide the current topology of physical and virtual networks.

3.1.4. *Virtualization hardware improvements*

The use of virtualized environments brings performance drawbacks particularly for network virtualization. The virtualization layer introduces overhead because it requires an extra task for VM multiplexing. This overhead is critical for I/O intensive loads [FER 10]. One of the critical cases of heavy I/O utilization is network intensive environments. In this case, providing high throughput and low delay are both required. Moreover, the network traffic of each VM must be isolated from each other.

Nowadays, I/O virtualization is under responsibility of the Virtual Machine Monitor (VMM), which must multiplex outgoing data flows and demultiplex incoming data flows. Regarding network virtualization, the VMM must share the link, controlling its access and multiplexing incoming packets to the correct virtual network interface. These tasks must be fair considering all VMs. To improve overall network virtualization performance, several techniques have been proposed: direct assignment of device [INT 12], multiqueued devices [INT 11b] and Single Root I/O Virtualization (SR-IOV) [INT 11a]. These techniques are briefly discussed in the following sections. We also discuss our current efforts on incorporating these new technologies in the prototype.

3.1.4.1. *New I/O virtualization techniques*

– *Direct I/O access*: The direct I/O technology is a new functionality provided by modern motherboard chipsets to safely allow direct device access from VMs. This technique allows device direct memory access (DMA) to different memory areas. Hence, it provides the ability for a device to transfer data directly to a VM, without VMM (or hypervisor) intervention. Memory accesses are assisted by the chipset of the motherboard, which intercepts device memory accesses and makes use of I/O page tables to verify whether the access is permitted and translate the required address to the physical memory address. This mechanism, however, has scalability problems because a physical device cannot be shared between several VMs. It can only be assigned to one VM.

– *Multiple queues*: The VMM (or hypervisor) has an important task of classifying packets. Packet classification incurs in great VMM processing overhead because it demands that the VMM define the destination VM of all incoming packets and multiplex all outgoing packets. Modern Network Interface Cards (NICs) address this problem by having multiple queues and doing packet classification by themselves. To accomplish this feature, the NIC classifies a packet using a certain pattern (virtual local area network (VLAN) tag or MAC destination address) and pushes it into the appropriate queue. One or more queues can be assigned to VMs. Thus, they have their traffic isolated from each other.

– *SR-IOV*: The SR-IOV standard allows sharing PCI-Express devices, such as NICs, among several VMs and accesses them as if they were native. The standard provides a way of bypassing the VMM involvement in data transfer.

This standard approach also defines a way of sharing an NIC to several VMs. NIC access, using multiple queues and direct I/O, follows the SR-IOV standard.

Xen version 4.0.0 has several new features related to the new I/O virtualization techniques. This version of Xen is used in the prototype. An important issue is the management domain kernel, the Domain 0 kernel. We need a compatible kernel supporting the new features to configure the devices and hypervisor, making the environment fully functional. The newest kernel that works with Xen 4.0.0 hypervisor is 2.6.32 Linux kernel (stable version). Several new features are still under development and are not currently supported. Currently, it is possible to give a VM full control of a device with Direct Access. It is also possible to create virtual functions, used to access PCI-Express devices, of SR-IOV specification, but it is not possible to give control of these virtual functions to a VM.

3.2. OpenFlow-based prototype

The OpenFlow-based prototype is based on a set of applications running over the **NOX!** (**NOX!**) controller [GUD 08]. NOX is an OpenFlow controller that configures and manages flows in OpenFlow switches. The developed NOX applications implement the five required primitives: instantiate flow (instantiate primitive), delete flow (delete primitive), migrate flow (migrate primitive), manage and change flows (set primitive), network monitoring and topology discovery (monitor primitive) and also a function to avoid loops in network. There is also a Web service interface to provide the five primitives to the piloting system.

The OpenFlow prototype architecture is illustrated in Figure 3.7. The prototype accomplishes both sensors and actuators defined by the primitives of the piloting system. The actuators are the Flow Manager and the Flow Migration applications. The Flow Manager application offers the instantiate/delete/set primitives. The Flow Manager application implements an interface between other NOX applications and the OpenFlow commands. This application is responsible for adding, modifying and deleting flows. The Flow Manager application receives a flow operation request and translates it into an OpenFlow command. The other actuator is the Flow Migration application, which implements the migration primitive. The Flow Migration migrates a flow from one path to other with no packet loss. The Stats and the Discovery applications implement the sensors in the prototype. Both applications satisfy the monitor primitive. The Stats application measures the network and collects statistics about switches. The Discovery application discovers the network topology and builds a graph to represent the network. There is also the Spanning Tree application that avoids the occurrence of loops in the network and unnecessary broadcast retransmissions.

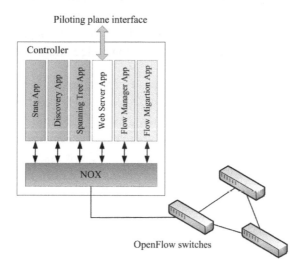

Figure 3.7. *The OpenFlow-based prototype architecture*

In addition, we use a tool that enables several NOX controllers to run in parallel in order to slice the network in several virtual networks. This tool, which is called FlowVisor [SHE 10], controls the network resource sharing, as bandwidth, topology seen by the controllers, the traffic of each share, the switch CPU usage and the flow table. FlowVisor implements the set primitive to configure the parameters of each virtual network.

The following sections describe the applications developed and modified for the network prototype. Section 3.2.1 presents the FlowVisor tool that allows network resource sharing among NOX controllers. Section 3.2.2 presents the Migration application, which is developed to migrate the flow path with no packet loss. Section 3.2.3 presents the Stats application that shows the statistics and information acquired from network and switches. Section 3.2.4 describes the Discovery application that discovers the network topology and builds a graph representation of the network. Finally, section 3.2.5 presents the Spanning Tree application that avoids loop occurrences and unnecessary broadcast messages in the network.

3.2.1. *FlowVisor*

The FlowVisor [SHE 10] is a special type of OpenFlow controller. It works as a transparent proxy between network devices and other controllers, such as NOX controllers. The main feature of FlowVisor is the ability to slice the network and share network resources in a controlled and isolated fashion.

As shown in Figure 3.8 (adapted from Sherwood *et al.* [SHE 10]), FlowVisor intercepts the OpenFlow messages sent by guest controllers. After that, FlowVisor, based on the user slicing policy (2), transparently modifies the message to delimit the control to a slice of the network. FlowVisor only forwards messages from switches to guests, if the messages match the slice policy. FlowVisor slices the network, keeping the slices isolated from each other.

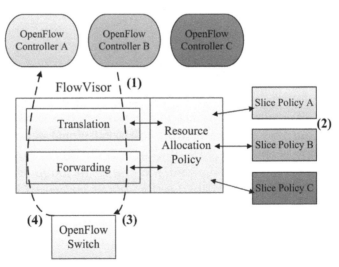

FlowVisor:
(1) Intercepts the OpenFlow messages sent by guest controllers.
(2) Verifies the user slicing policy.
(3) Modifies the message to delimit the control to a slice of the network.
(4) Forwards messages from switches to guests, if the messages match the slice policy.

Figure 3.8. *The FlowVisor operation*

FlowVisor can virtualize the network by slicing resources of switches among several controllers. Additionally, FlowVisor allows the creation of hierarchies of FlowVisors, modifying the network architecture, as shown in Figure 3.9 (adapted from Sherwood *et al.* [SHE 10]). In this case, one FlowVisor instance (FlowVisor 2) virtualizes several switches in parallel (Switches 1 to 4), and FlowVisor 1 recursively slices the virtual slices defined by FlowVisors 2 and 3. To slice the network among controllers, FlowVisor focuses the sharing on five network resources: bandwidth isolation, topology discovery, traffic engineering, device CPU monitoring and forwarding tables control. For example, bandwidth isolation is ensured by using different priority queues on switches. One queue is isolated from the others. All packets in a flow are marked with the same identifier and then are mapped onto one of eight priority queues. Flows can be classified based on three different identifiers: VLAN Priority Code Point (PCP), IP Type of Service (ToS) and OpenFlow Quality-of-Service (QoS). The default is the VLAN PCP. It is important to note that what is guaranteed is the minimum bandwidth.

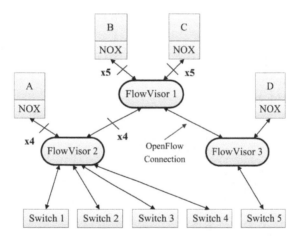

Figure 3.9. *The FlowVisor hierarchy: FlowVisor 1 recursively slices the virtual slices defined by the FlowVisors 2 and 3*

3.2.2. *OpenFlow migration*

Flow Migration is a NOX [GUD 08] application developed for the prototype that defines a new path between source and destination OpenFlow switches, and changes the current path of the flow to the new path with no packet loss during this procedure. To migrate a flow, the Flow Migration application takes, as parameters, a list of switches. This list defines an ordered group of switches through which the new path flow must pass. To calculate the shortest path from the source to the destination including the selected switches, it uses a generalization of the Dijkstra Algorithm [MOY 98].

The Flow Migration application has two interfaces: one with the Web server application, by which the control parameters are changed, and the switch interface, where the application sends the flow configuration commands. Figure 3.10 illustrates an example of the flow migration procedure. Initially, the source sends packets to the destination through the Path 1, composed by Nodes A-F-E-D. A new flow from Switch A to D is defined, with the requirement that it must pass through the Switches A, B and D, the dark gray nodes. A, B and D are not a complete path from A to D. Thus, the Flow Migration application calculates a complete path from source to destination switches using the Dijkstra algorithm to minimize the number of hops in the new path. According to the algorithm, Node C must be included in the path. After defining all nodes that compose the entire path, the application configures the flow in the switches in the reverse direction, that is it configures the flow from the switch closest to the destination to the farthest switch. Because all paths from destination to source are already configured but the link of the source computer to the Switch A, which is the last link to be changed, avoids packet loss.

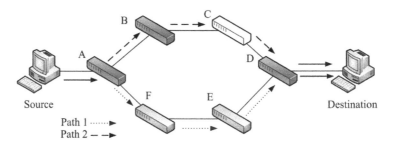

Figure 3.10. *The operation of Flow Migration application*

3.2.3. *OpenFlow statistics*

The Stats application is another NOX application developed for gathering OpenFlow switches information and converting them into XML messages. The Stats application sends commands to the OpenFlow switched network, querying each switch about its statistics and other information. Each switch receives a request to report its description, datapath, table, aggregated flows and flow information to NOX controller. Each kind of information is obtained with an appropriate OpenFlow Protocol command. The Stats application interfaces with the Web server and Switch Stats applications. The Web server application provides an interface between the Stats application and a HyperText Transfer Protocol (HTTP) client. The client can be a user interface or another application. The Switch Stats application is responsible for sending OpenFlow commands to switches and handling the responses. For simplicity, we will refer to Switch Stats application and Stats application, both together, as Stats application.

Stats application periodically gathers information from the OpenFlow switched network. Whenever a stats request is received, the Stats application responds with an XML message with the currently available information. The XML message returned by the Stats application contains the information of every switch registered on the NOX controller. The XML message begins with a root tag named openflowstats and has a datapath tag for every registered switch. Inside every datapath tag, there is the switch identifier number (dpid), the switch MAC address (dp_mac) and specific tags containing information and statistics retrieved from the switch. These specific tags and internal structures store switch information and statistics such as switch IP address (ip), hardware description (hw_desc), the number of switch implemented tables (n_tables), sent and received packets, flows count (flow_count) and flow priority (priority).

3.2.4. *OpenFlow discovery*

The Discovery application is a modified version of the default NOX application. First, we have made this application compatible with any OpenFlow device and we have also developed an interface for the Web server application. The Discovery application also gets the topology of both physical and virtual networks.

Basically, the Discovery application implements the Link Layer Discovery Protocol (LLDP) [IEE 05]. This protocol allows nodes to transmit information about the capabilities and the current status of network devices. The LLDP implementation is optional in the protocol stack of IEEE 802 LAN stations. The LLDP frame format is illustrated in Figure 3.11. The LLDP frame has the device information stored in Type-Length-Value (TLV) structures. The Discovery application then creates an LLDP frame for all the ports of a given switch. The switch propagates those frames through its ports. Upon receiving the first LLDP frame, the switch does not know the forwarding rules for that frame. The switch forwards the frame for the controller to analyze its content. After receiving the frame the controller requests the running instance of the Discovery application to process this frame. Thus, the application analyzes which switch received the frame, from which port the frame was received, which switch sent the frame, and in which port the frame was sent. With this information, the application creates a data structure with source switch, source port, destination switch and destination port. This data structure identifies a link and the aggregation of all links of a network characterizes the network topology.

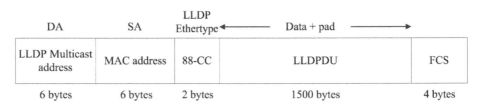

Figure 3.11. *LLDP frame format for IEEE 802.3*

The algorithm of the default Discovery NOX application is modified to guarantee the correctness of the topology discovery procedure on Type-1 OpenFlow switches. On Type-0 switches, the simplest model of OpenFlow switches, a frame is forwarded only if the controller has already configured a flow for that frame. If there is no configured flow, the frame is dropped. However, Type-1 switches process the traffic of a flow even if there is no specific rule configured by the controller, that is, by default, initial packets are not dropped automatically. This fact leads to an incorrect representation of the network topology.

Suppose an LLDP sent from Switch A to Switch B. With the default algorithm, the controller will not send the command to drop this frame after processing it. In this case, B may incorrectly forward this LLDP frame to other switches connected to it. Suppose that Switch B and C are directly connected but C is not directly connected to A. In this case, B forwards the LLDP frame to C and thus the controller will wrongly identify a link between

A and C. Consequently, a wrong topology is assumed. To fix this problem, we modify the default algorithm in order to allow it to always send the drop command to the switches. The differences between the two algorithms are represented in Figure 3.12.

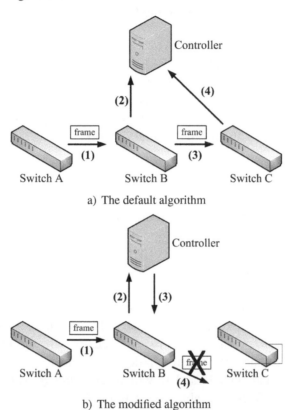

a) The default algorithm

b) The modified algorithm

Figure 3.12. *An example of the behavior of default and modified Discovery algorithms: the first two steps of both mechanisms are the same but with the modified algorithm the controller sends the drop command to Switch B that drops the LLDP frame instead of forwarding it to Switch C*

Figure 3.12a) represents the default discovery algorithm. All switches are Type-1. In this example, (1) Switch A sends a LLDP frame to Switch B. After receiving this frame, (2) B sends it to the controller. The controller then identifies that there is a link between A and B. The controller does not send a drop command to B that (3) forwards the LLDP frame to C. Switch C receives the frame, and consequently (4) sends it to the controller. Then, the controller incorrectly assumes that there is a link between A and C. The modified version of the algorithm is illustrated in Figure 3.12b). The first two steps are the same as that of the default algorithm. The difference is that, after this, (3) the controller sends the drop command to B and thus (4) B drops the LLDP frame instead of forwarding it to C.

The default application only provides the physical network topology and we also need information about virtual networks in the prototype. To achieve this goal, the modified Discovery application receives the definition of the virtual network, for instance a VLAN identifier or an IP range, and then it provides a structure with the virtual network topology, which comprises all the switches that are forwarding traffic for that virtual network.

3.2.5. *OpenFlow spanning tree*

Multiple paths between a source-destination pair is a common characteristic of different network topologies. These redundant paths increase network reliability but can cause errors during broadcast procedures in Ethernet networks. In fact, if a node connected to a switch broadcasts a frame, the switch forwards this frame to all ports but the incoming port. If the network has redundant paths, the frame will be broadcast again by all switches until it returns to the source switch. For example, if Node B sends a broadcast message, A and C will receive and forward the message. When Node A receives the message forwarded by Node C, it will forward the message again, and both B and C will receive and forward the message. Because Switch B is not able to identify that the frame was already forwarded, it forwards the frame again. Hence, the frame will be continuously forwarded due to a loop in the network topology. We use the Spanning Tree algorithm [GIB 11] to control the physical topology and to avoid these loops on our OpenFlow prototype. This algorithm creates a topology for broadcasting frames without creating loops.

The Spanning Tree algorithm runs as a NOX application. The algorithm builds a graph that maps the network topology. Each node and each edge of the graph represents a switch and a link, respectively. The algorithm analyzes the graph and builds a tree containing all the nodes of the graph and the selected edges, as shown in Figure 3.13. Disabling all links (edges) outside of the spanning tree means these disabled links will not forward broadcast messages. This mechanism prevents frames from traversing disabled links, loop occurrence and unnecessary broadcast traffic retransmission.

The Spanning Tree algorithm uses topology information provided by the Discover application and works as follows. First, the algorithm sorts the switches (nodes) ordered by their identifiers (Ids). After that, it selects the switch that has the lowest identifier as the root of the spanning tree. Then, the Spanning Tree algorithm checks the switches connected to the root and marks these switches as visited, which means that the node and the link between these nodes and the root also compose the spanning tree. Finally, it checks the switches connected to the already-visited switches. If any of these switches is marked as visited, the link between these switches does not compose the spanning tree. Otherwise, the algorithm inserts both the node and the corresponding link in the spanning tree. It uses a breadth first search algorithm [ZHO 06]. If we start at a particular node (say, $n1$), the breadth first search algorithm searches all nodes $M - hops$ away from $n1$ before all nodes $M + 1 - hops$ away from $n1$. For instance, in Figure 3.13, if $n1 == B$, then the algorithm searches D and E before G and H.

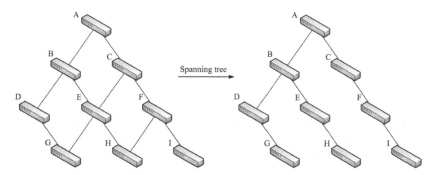

Figure 3.13. *A network topology and its corresponding spanning tree, assuming A as the root of the spanning tree*

The Spanning Tree algorithm runs periodically to keep the spanning tree updated. The insertion of a new switch to the controller demands that the Spanning Tree disable all ports of this switch until it analyzes the graph and includes the new switch in the spanning tree. For security, after a minimum amount of time, the Spanning Tree application enables these ports to ensure that they will be identified by the Discovery application, and, consequently, included in the network topology.

When a switch receives a new broadcast data flow, it forwards the frame to the controller. After receiving the frame, the controller calls the running instance of the Spanning Tree to analyze the link through which the broadcast frame was received. If this link is in the spanning tree, the switch processes and forwards the frame. Otherwise, the application drops the frame. When the frame received is an LLDP frame, the processing and forwarding occurs independently of the source link because the Discover application uses that frame to discover the network topology. Also, if the controller has already defined a rule in the switch for the received broadcasted frame, the switch does not forward that frame to the controller, but processes the frame according to the defined rule instead, even if the link is not in the spanning tree.

3.3. Summary

This chapter has described the required improvements on two virtualization tools, Xen and OpenFlow, to implement the five primitives (instantiate, delete, migrate, monitor and set) used by the piloting plane. Two network prototypes were described. Both are composed of network elements able to sense and to act on the virtual environment.

In the Xen-based prototype, modules that gather information about the environment, such as the Xentop Gatherer, the Ifconfig Gatherer, the Latency Gatherer and the Memory Gatherer, are the sensors. There is also the Xen Topology module that allows the discovery of physical and virtual network topologies. In the OpenFlow-based prototype, the sensor capability relies on the developed NOX applications. The Stats application allows the controller to retrieve information about the status of each OpenFlow switch, such as flows that are traversing the switch and number of packets received per flow. The Discovery application discovers the network topology and the Spanning Tree application avoids the unnecessary broadcast traffic retransmission and the occurrence of network loops. The actuating capability allows the controller to actuate on the network and perform tasks such as modifying element configurations on the fly and changing a virtual topology with the aid of the *migrate* primitive. In the Xen-based prototype, a modified migration tool allows dynamic reconfiguration of virtual topologies with no packet loss. In the OpenFlow-based prototype, the Flow Migration application can modify a flow path with no packet loss as well. Through these actuators, the piloting plane can dynamically reorganize network topology in order to ensure users' requirements. These modules and applications developed for both prototypes are the basis for the development of the piloting plane, which is discussed in Chapter 4.

3.4. Bibliography

[ALK 10] ALKMIM G.P., DA FONSECA N.L.S., "Virtual network mapping on a physical substrate", *Workshop on Network Virtualization and Intelligence for Future Internet (WNetVirt)*, Búzios, RJ, Brazil, April 2010.

[ALV 12] ALVES R.S., CAMPISTA M.E.M., COSTA L.H.M.K., *et al.*, "Towards a pluralist Internet using a virtual machine server for network customization", in *Asian Internet Engineering Conference (AINTEC'2012)*, Bangkok, Thailand, pp. 9–16, November 2012.

[CLA 05] CLARK C., FRASER K., HAND S., *et al.*, "Live migration of virtual machines", *Usenix Symposium on Networked Systems Design and Implementation (NSDI)*, Boston, MA, USA, pp. 273–286, May 2005.

[FER 10] FERRAZ L., CARVALHO H., PISA P., *et al.*, "New I/O virtualization techniques", *Workshop on Network Virtualization and Intelligence for Future Internet (WNetVirt)*, Búzios, RJ, Brazil, April 2010.

[FRE 10] FREITAS R.B., DE PAULA L.B., MADEIRA E., *et al.*, "Using virtual topologies to manage inter-domain qos in next generation networks", *International Journal of Network Management*, vol. 20, no. 3, pp.111–128, 2010.

[GIB 11] GIBB G., "Basic spanning tree". Available at http://www.openflowswitch.org/wk/index.php/Basic_Spanning_Tree, 2011 (accesssed in May 2013).

[GUD 08] GUDE N., KOPONEN T., PETTIT J., *et al.*, "NOX: towards an operating system for networks", *ACM SIGCOMM Computer Communication Review*, vol. 38, pp.105–110, July 2008.

[IEE 05] IEEE, 802.1ab IEEE standard for local and metropolitan area networks. Station and media access control connectivity discovery, Technical report, IEEE Institute of Electrical and Electronics Engineers, 2005.

[INT 11a] INTEL LAN ACCESS DIVISION, PCI-SIG SR-IOV Primer: An Introduction to SR-IOV Technology, 2011, available at http://www.intel.com/content/dam/doc/application-note/pci-sig-sr-iov-primer-sr-iov-technology-paper.pdf, (accessed in May 2–13).

[INT 11b] INTEL CORPORATION, *Intel 82576 Gigabit Ethernet Controller Datasheet*, October 2011. Available at http://www.intel.com/content/dam/www/public/us/en/documents/datasheets/82576eb-gigabit-ethernet-controller-datasheet.pdf, (accessed in May 2013).

[INT 12] INTEL CORPORATION, Intel Virtualization Technology for Directed I/O (VT-d): Enhancing Intel platforms for efficient virtualization of I/O devices, 2012 available at http://software.intel.com/en-us/articles/intel-virtualization-technology-for-directed-io-vt-d-enhancing-intel-platforms-for-efficient-virtualization-of-io-devices (accessed in May 2013).

[KAR 10] KAROUIA A., LANGAR R., NGUYEN T.-M.-T., *et al.*, "SOA-based approach for the design of the future Internet", *Communication Networks and Services Research Conference (CNSR)*, Montreal, Canada, pp. 361–368, May 2010.

[MAT 11] MATTOS D.M.F., FERNANDES N.C., DA COSTA V.T., *et al.*, "OMNI: OpenFlow MaNagement Infrastructure", in *2nd IFIP International Conference Network of the Future (NoF'2011)*, Paris, France, November 2011.

[MOY 98] MOY J., "OSPF version 2", IETF Network Working Group RFC 2328, April 1998.

[PIS 10] PISA P.S., FERNANDES N.C., CARVALHO H.E.T., *et al.*, "OpenFlow and Xen-based virtual network migration", *World Computer Congress – Network of the Future Conference*, Brisbane, Australia, pp. 170–181, September 2010.

[PIS 11] PISA P.S., COUTO R.S., CARVALHO H.E.T., *et al.*, "VNEXT: Virtual NEtwork management for Xen-based Testbeds", *2nd IFIP International Conference Network of the Future - (NoF'2011)*, Paris, France, November 2011.

[SEN 10] SENNA C.R., BATISTA D.M., MADEIRA E.R.M., *et al.*, "Experiments with virtual network management based on ontology", *Workshop on Network Virtualization and Intelligence for Future Internet (WNetVirt)*, Búzios, RJ, Brazil, April 2010..

[SHE 10] SHERWOOD R., CHAN M., COVINGTON A., *et al.*, "Carving research slices out of your production networks with OpenFlow", *ACM SIGCOMM Computer Communication Review*, vol. 40, no. 1, Búzios, RJ, Brazil, pp. 129–130, April 2010..

[SOA 10] SOARES JR. M.A., MADEIRA E.R.M., "Autonomic management of resources in virtualized networks", *Workshop on Network Virtualization and Intelligence for Future Internet (WNetVirt)*, Búzios, RJ, Brazil, April 2010.

[WOL 02] WOLFGANG M.N., "Host discovery with nmap", November 2002, Available at http://www.nmap.org (accessed in May 2013).

[VAN 10] VANT HOOFT F.N.C., MADEIRA E.R.M., "Resource allocation policies in future multi-agent based virtual network", *Workshop on Network Virtualization and Intelligence for Future Internet (WNetVirt)*, Búzios, RJ, Brazil, April 2010..

[WAN 08] WANG Y., KELLER E., BISKEBORN B., *et al.*, "Virtual routers on the move: live router migration as a networkmanagement primitive", *ACM Special Interest Group on Data Communication Conference (ACM SIGCOMM 08)*, Seattle, Washington, USA, pp. 231–242, 2008.

[ZHO 06] ZHOU R., HANSEN E.A., "Breadth-first heuristic search", *Artificial Intelligence*, vol. 170, nos. 4–5, pp. 385–408, 2006.

Chapter 4

State of the Art in Context-Aware Technologies

The goal of the Horizon Project is to conceive and test a new architecture for a post-Internet Protocol (IP) environment. This post-IP architecture employs network virtualization, which includes a piloting system. The design of such a pilot system is based on multi-agent systems (MASs) to guide the system toward intelligent decisions.

The main idea of this project is to develop an environment where each element of the system contributes with information used to automatically update the control algorithms to reflect the changes in the environment that influence the value of the network parameters. For that, the piloting system must be "autonomous". To make decisions for enabling the system with such autonomy, we surveyed the state of the art on autonomous systems, MASs with support to piloting systems as well as some development platforms that offer such features.

This chapter describes the state of the art on MASs and piloting systems. Section 4.1 describes the characteristics of autonomous systems. Section 4.2 introduces MASs with emphasis on their specific features to support piloting systems. Section 4.3 describes agent platforms that fulfill the requirements of the project, and section 4.5 concludes the chapter.

Chapter written by Edmundo R.M. MADEIRA and Guy PUJOLLE.

4.1. Autonomic systems

An autonomic computing system can be described as a system that senses its operating environment, and models its behavior on that environment. Moreover, it takes actions to change either the environment or its own behavior. A goal-oriented system is an autonomic system that operates independently and that achieves its goals by itself without intervention, even if external environmental changes. An autonomic system comprises the properties of self-configuration, self-healing, self-optimization and self-protection [IBM 06]. In summary, the main characteristics of autonomic systems are autonomy and spontaneity.

4.1.1. *Characteristics of autonomic systems*

The concept of "The Autonomic Computing" was proposed by IBM [IBM 06], which defined four primary functionalities of autonomic computing: self-configuration, self-healing, self-optimization, and self-protection, which are described as follows:

– *Self-configuration* is the capacity of adapting itself to dynamically changing environments. When a stand-alone component is introduced, it integrates seamlessly into its surrounding environment and the rest of the system automatically adapts itself to the presence of the new components. The system has the capacity to self-adjust, and the automatic configuration of components follows high-level policies.

– *Self-healing* is the ability to discover, diagnose and act to prevent disruptions. The system must be able to maintain all its features, possibly in a degraded mode, until all the needed resources be found. It should maintain a base of knowledge about the configuration system, to allow diagnostic reasoning and to analyze system logs (or logs from other systems) to identify failure.

– *Self-optimization* is the capacity to tune resources and to balance workload so that the use of resources can be measured. In this way, components should be structured to improve performance and efficiency, which demands the ability to monitor the environment, experiment with new options and learn to improve choices for performance optimization.

– *Self-protection* is the ability to anticipate, detect, identify and protect against threats. An autonomous system must detect these situations and avoid disruption of usage. Such ability calls for the setting up of mechanisms and architecture for detection and protection of all network elements.

4.1.2. *Architecture and operation of autonomic systems*

An autonomic computing system is composed of a connected set of autonomic elements. Each element must include sensors and actuators [STE 03]. The system goal is achieved by a control loop composed of four main functions: system behavior monitoring by acquisition of sensor measurements, comparison of sensor measurements with expectations, decision making if the system behavior is not the expected one, and execution of actions to change the system behavior.

4.1.2.1. *Architecture of autonomic elements*

Figure 4.1 shows an architecture of an autonomic element [STE 03].

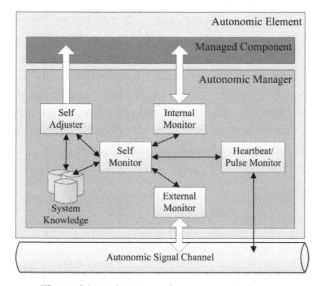

Figure 4.1. *Architecture of an autonomic element*

An autonomic element is composed of managed components, as well as an autonomic manager that contains several components:

– *The internal monitor* observes the state of the "managed component" and communicates the information to the self-monitor component.

– *The external monitor* observes the state of the environment and communicates this state information from environment to the self-monitor component.

– *The self-monitor* assesses the state of the managed component, compares it with the expected state stored in a knowledge base, obtaining then the deviation of the expected state of managed component.

– *The self-adjuster* allows the adjustment of the status of the managed component.

– *The heartbeat monitor* summarizes the state of the autonomic entity and communicates with other entities responsible for the autonomic state control.

Figure 4.1 shows that an autonomic element is composed of a managed component and a correspondent autonomic manager. The autonomic manager implements the required self-monitoring and self-adjusting capabilities. An internal monitor observes the state of the managed component and passes this information to the self-monitor for evaluation and action. The measured state is compared with the expected state held in a knowledge base. Undesirable deviations are reported to the self-adjuster for action, which may result in changes to the managed component. Similarly, an external monitor observes the state of the environment via an autonomic signal channel and it may trigger internal changes. The signal channel links it to other autonomic managers. The heartbeat or pulse monitor provides a summary of the state of an autonomic element to other autonomic elements responsible for monitoring that state [STE 03].

In summary, an autonomic system composed of a set of autonomic entities operates in a control loop that ensures the "self" properties.

4.1.2.2. *Autonomic control loop*

Autonomic systems work as control loops (Figure 4.2) [DOB 06]. These systems collect information from a variety of sources including traditional network sensors and reporting streams as well as higher level devices and user context. The information is analyzed to construct a model of the evolving state of the network and its services, so that adaptation decisions can be made. These decisions are actuated through the network and will potentially be reported to users or administrators. The impact of these decisions can then be collected to inform the next control cycle.

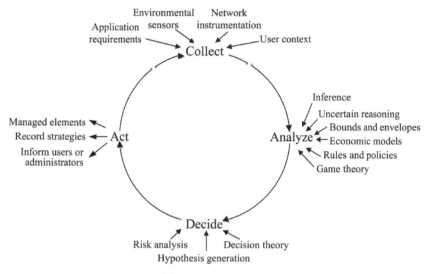

Figure 4.2. *Autonomic control loop*

Several solutions have been proposed for the implementation of these systems, including intelligent agents. In the following section, we present information about MASs.

4.2. Piloting with multi-agent systems

The correspondence between a population of cooperating autonomous agents and the characteristics of telecommunication networks is the main motivation for using an MAS in telecommunication networks. Researchers on MASs have searched for challenging study cases, while the network community searches new solutions to some of the traditional problems, such as provisioning of Quality-of-Service (QoS) and avoidance of manual network configuration [BUL 08].

Agents are usually classified in two categories: "cognitive" and "reactive"; the difference between them involves the design methods and their usage. Ferber [FER 95] summarizes such difference by the question: should we understand agents as entities already intelligent that are capable of solving certain problems by themselves, or should we assimilate them as very simple reactive beings that act directly to environmental changes?

4.2.1. *Definition of agents*

There is not a universally accepted definition of the concept of "agent". One of the proposed definitions [FER 95, WOO 02] is that an agent is a physical or virtual entity which:

– is capable of acting in an environment;

– can communicate directly with other agents;

– is driven by a set of trends (in terms of its objectives, or a function of satisfaction and even survival, that it seeks to optimize);

– possesses its own resources;

– is capable of perceiving (in a limited way) its environment;

– has only a partial representation of this environment (and possibly even none);

– possesses skills and offers services;

– can possibly reproduce itself;

– tending to achieve its objectives, by taking into account the available resources and skills. It perceives, represents and communicates with other agents.

Thus, an agent can be seen as an entity capable of thinking and acting independently of its environment in order to satisfy objectives that were preset (by itself or by an external entity). Some characteristics of agents are as follows:

– *Ubiquity*, which is the capacity of sophistication and deployment of an agent-based process.

– *Interconnection*, which plays an essential role in the design of MASs.

– *Intelligence*, which can be measured by the complexity of tasks that are automated or delegated without human intervention.

4.2.2. *Characteristics of agents*

An agent can be situated, autonomous, proactive, reactive or social, as described below:

– *Situated*: the agent is able to act on its environment based on sensory inputs it receives from the environment.

– *Autonomous*: an agent should be able to act without the intervention of others (human or agent) and control its own actions and its internal state.

– *Proactive*: an agent must exhibit proactive behavior, while being able to take initiative at the right moments.

– *Reactive*: an agent must be able to perceive its environment and to formulate responses in the required time.

– *Social*: an agent must be able to interact with other agents (software or human) to perform its tasks or to assist other agents in performing their tasks.

4.2.3. *Cognitive agents*

There is no unique definition of "cognitive agents". In [FER 95], there are references to a "cognitive school". Researchers that follow this "school" are interested in agents that can make plans for their behaviors. A cognitive agent has a knowledge base including all information and know-how necessary to carry out its task and the interactions with other agents and with its environment. In other words, cognitive agents can be defined as "intentional". They possess goals and explicit plans to accomplish these goals. Briot [BRI 01] discusses deliberative agents, which are equivalent to the cognitive agents defined in [FER 95].

4.2.4. *Reactive agents*

While cognitive agents can build plans for their behaviors, the reactive agents just have reflexes. Reactive agents are defined as a special type of agent, cognitive agents being the general case. In [WOO 02], reactive agents are defined as those that react without reference to their history. In other words, "they simply respond directly to their environment". In [BRI 01], the authors describe the architecture of reactive agents as opposed to cognitive agents. Similarly to the architecture proposed in [WOO 02], the architecture discussed in [BRI 01] defines reactive agents as those that make decisions based only on the information collected in the current execution time.

4.2.5. *Multi-agent systems*

An MAS consists of a set of computer processes acting at a certain time, sharing common resources and communicating with each other. The key point of MASs is the formalization of coordination among agents. The main issues involving agents are as follows:

– *Decisions*: what are the mechanisms of the officer's decision? What is the relationship between perceptions, representations and actions of agents? How do they break down their goals and tasks? How do they construct representations?

– *Control*: what are the relationships between agents? How are they coordinated? This coordination can be described as cooperation to accomplish a common task or as a negotiation between agents with different interests.

– *Communication*: what kind of message do they send to each other? Which syntax do these messages follow? Different protocols are offered depending on the type of coordination between agents.

The MASs have applications in the field of artificial intelligence, which may reduce the complexity of the problem dividing it into subgroups. An intelligent agent is associated with each subgroup and the exchange of information between agents is used to perform coordination [FER 95]. This is known as distributed artificial intelligence. Multi-agents are especially useful in telecommunications, such as in electronic commerce, but can also be used in other applications such as optimization of transportation systems and robotics.

Ferber [FER 95] defines an MAS (Figure 4.3) as being composed of the following:

– *An environment E.*

– *A set of objects O* that are passive and can be associated with a position in environment E. They can be created, collected, modified and destroyed by agents.

– *A set A of agents* that represent entities active in the system.

– *A set R of relations* that bind objects (and therefore agents) between them.

– *A set O of operations* that allows which agents of the set A are able to collect, produce, consume, transform and manipulate objects from O.

– *Operators* which represent the application of the operations and the world's reaction to this attemp of change.

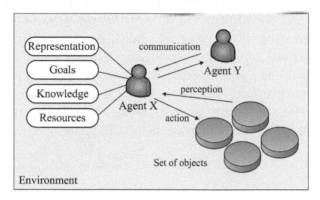

Figure 4.3. *A multi-agent system architecture: agents are able to perceive and act on objects of a given environment and interact with each other to act based on their goals and motivations*

An MAS is therefore composed of several agents that operate in an environment and interact with each other. These agents are able to perceive and act on objects detectable in this environment. Their collaboration allows them to act according to their respective goals and motivations. An MAS is composed of a set of agents that interact among themselves. Each agent is composed of the following components:

– *Collection component*, which allows agents to collect information on their environment.

– *Interaction component*, which allows agents to interact with other agents.

– *Decision component*, which allows agents to make decisions based on their perceptions.

– *Execution component*, which allows agents to perform their decisions.

4.3. Options to build the autonomic platform

Future post-IP architecture should be context aware. Then, the platform to accomplish such functionality includes physical and logical sensors (software entity, network components and software agents) to collect context information related to the presence, location, identity and profile of users and services. Typical context aware software involves the localization of services and users, the call of services according to user behavior, the provisioning of information for service composition, the facilitation of *ad hoc* communication among users and the adaptation of QoS to the changing environment. The objective is to explore the types of context-aware infrastructures and to choose the best way to introduce intelligence in the platform composed of a knowledge plane and a piloting plane. In this section, we describe what could be an autonomic platform. For this, we analyzed three platforms that could be useful in the Horizon Project: the Ginkgo platform, the Dimax platform and the Java Agent Development Framework (JADE) platform.

4.3.1. *Ginkgo*

The Ginkgo's technology provides support to autonomic networking applications by employing intelligent agents, distributed in network elements (NE) across the network. Ginkgo intelligent agents play a dual role in autonomic networking applications: by providing, in real time, the knowledge plane with the information required by the application and by using the distributed knowledge in the knowledge plane to manage, in real time, the network control mechanisms [GIN 08].

4.3.1.1. *Situated view of the Ginkgo agents and knowledge plane*

In the Ginkgo model, the knowledge plane is distributed among Ginkgo agents as a set of "situated views". Each situated view represents the knowledge of an individual agent regarding the situation of the network in its neighborhood, which is supposedly more important to the agent than the situation in remote locations. Furthermore, updating knowledge in individual situated views can be more easily done in real time with limited amounts of control traffic. Ginkgo agents working together maintain an always-up-to-date collective distributed knowledge of the overall network situation, as shown in Figure 4.4 [GIN 08].

4.3.1.2. *The Ginkgo agent architecture*

Ginkgo agents are made up of three main types of building blocks (Figure 4.5):

– *The situated view knowledge base*, which is dedicated to storing the structured knowledge of the situated view of the agents.

– *The behaviors*, which are autonomic software components permanently adapting themselves to the environmental changes. Each of these behaviors can be considered as a specialized function with expert capabilities. Each behavior is essentially a sense→decide→act loop in charge of a control function.

– *The dynamic planner*, which is in charge of orchestrating the behaviors. The dynamic planner follows a policy provided by the user indicating how the behaviors should take into account the dynamic changes occurring in the environment.

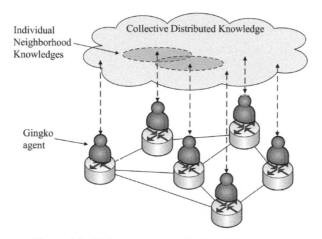

Figure 4.4. *Ginkgo agents and the knowledge plane*

Each of these behaviors ("B" in Figure 4.5) can be executing some expert function. Typical functions of behaviors are as follows:

– Production of knowledge for the situated view in cooperation with other agents.

– Reasoning individually or collectively to evaluate the situation and to decide on the application of an appropriate action, for example a behavior can simply be in charge of computing the available bandwidth on the network element (NE). It can also regularly perform complex diagnoses or it can be associated with automatic recognition of specific network conditions.

– Action onto the NE parameters, for example a behavior can tune QoS parameters in a DiffServ context.

– Uploading of useful information to a network management system, for example a behavior can upload a synthetic alarm obtained from the observations of elementary alarms.

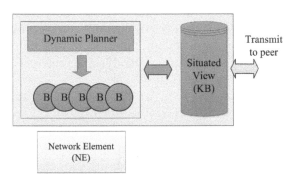

Figure 4.5. *Ginkgo agents architecture*

Behaviors have access to the situated view, which operates within each agent as a whiteboard shared among the behaviors of the agent.

The activation, dynamic parametrization and scheduling of behaviors within an agent is performed by the dynamic planner. The dynamic planner decides which behaviors have to be active, when they have to be active and with which parameters. The dynamic planner detects changes in the situated view and occurrence of external/internal events. It orchestrates the reaction of the agent to change the network environment. To do so, the dynamic planner follows a rule-based policy expressed in a simple and compact form.

The situated view of each agent is a structured knowledge base representing the environment of the agent. It contains knowledge elements collected locally by the agent as well as knowledge obtained from its peers. The situated view is updated on a periodical basis and it is used to adapt the behaviors to changes occurring in the network and to take real-time decisions. An automatic mechanism mirrors the situated view to the appropriate peers; the knowledge is reflected in the situated view of the peer agents. The rate and range of this mechanism can be tuned according to the nature of the knowledge. The situated view is organized following an ontology-based model, which helps to build well-structured applications as well as to interoperate with other systems.

4.3.2. *DimaX*

DimaX is a fault-tolerant multi-agent platform that offers several services like naming, fault detection and recovery. To make MAS robust, DimaX uses replication techniques. Moreover, DimaX provides developers with libraries of reusable components for building MAS. DimaX presents interesting features such as robustness and reusability.

4.3.2.1. *DimaX services*

DimaX [FAC 06] is the result of the integration of a multi-agent platform (named DIMA) and a fault tolerance framework (named DARX). Figure 4.6 gives an overview of DimaX and its main components and services. DimaX is designed in three levels: system (i.e. DARX middleware), application (i.e. agents) and monitoring. At the application level, DIMA provides a set of libraries to build multi-agent applications. Moreover, DARX provides the mechanisms necessary for distributing and replicating agents as services. DimaX server provides the following services: naming, fault detection, observation and replication.

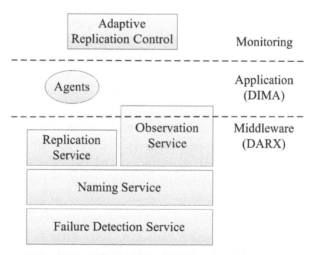

Figure 4.6. *DimaX architecture levels: middleware,*
application and monitoring

The *Naming service* maintains a list (i.e. white pages) of all the agents within its administration domain. When an agent is created, it is registered at both the DimaX server and the naming server. The *failure detection service*

(from DARX) is based on the heartbeat technique; a process sends an I am
alive message to other processes to inform them that it is safe. When a server
detects a failure of another DimaX server, its naming module removes all the
replicated agents hosted at the faulty server from the list and replaces them by
their replicas located on other hosts. The replacement is initiated by the failure
notification.

The functionalities of the *observation service* are fundamental for
controlling replication. An observation module collects data at two levels.
The system level collects data about the execution environment of the MAS,
such as CPU time and mean time between failures, while the application level
collects information about its dynamic characteristics, such as the interaction
events among agents (e.g. the sent and received messages). The observation
service relies on a reactive agent's organization (named host monitors). These
agents collect and process the observation data to compute local information,
such as the number of exchanged messages between two agents during a
given period.

DimaX uses replication mechanisms (*replication service*) to avoid failures
of MASs. The replication service enables running MASs without interruption,
regardless of the failures. A replicated agent is an entity that possesses two or
more copies of its behavior (or replicas) in different hosts. There are two main
types of replication protocols: active and passive. In active replication, all
replicas process all input messages concurrently, while in passive replication,
only one of the replicas processes all input messages and periodically
transmits its current state to the other replicas so that consistencies are
maintained. DimaX offers several libraries and mechanisms to facilitate the
design and implementation of fault-tolerant MASs. The characteristics of
DimaX agents are briefly described in the following.

4.3.2.2. *DIMA agent behaviors*

DIMA is a Java multi-agent platform. Its kernel is a framework of
proactive components that represent autonomous and proactive entities. A
simple DIMA agent architecture consists of a proactive component, an agent
engine and a communication component. The proactive component (the
AgentBehavior class) represents the agent behavior. This proactive
component includes a decision component to select appropriate actions. For
instance, a finite state machine or a rule-based system could be used to
describe the decision process. The selected actions can include sending

messages and a communication component sends and delivers messages. An agent engine is provided to launch and support the agent activity.

4.3.2.3. DarX tasks

DARX is a framework for designing reliable distributed applications, which include a set of distributed communicating entities (named DarX tasks). It includes transparent replication management. While the application deals with tasks, DARX handles replication groups. Each of these groups consists of software entities (the replicas), which represent the same DarX task. In DARX, a DarX task can be replicated several times and with different replication strategies.

4.3.2.4. Fault-tolerant agents

A fault-tolerant agent (called DimaX agent) is an agent built on our DimaX fault-tolerant multi-agent platform. Each DimaX agent has the structure of a DarX task. The DarX task, however, is not autonomous. To make it autonomous, we encapsulate the DIMA agent behavior in it. This agent architecture enables the replication of the agent. As the DARX middleware and the DIMA platform provide mechanisms for execution control, communication and naming at different levels, their integration requires a set of additional components. These components set calls, transparently, DARX services (e.g. replication and naming), when executing multi-agent applications developed with DIMA. At the application level, any code modification is required. It controls the execution of agents built under DimaX and it offers a communication interface between remote agents, through DimaX servers.

4.3.3. JADE

JADE is a software environment for building agent systems for the management of networked information resources in compliance with the Foundation for Intelligent Physical Agents (FIPA) specifications for interoperable MASs [FIP 10]. JADE provides a middleware for the development and execution of agent-based applications, which can work and interoperate seamlessly both in wired and wireless environments. Moreover, JADE supports the development of MASs based on a predefined programmable and extensible agent model as well as on a set of management and testing tools [BOR 05]. A JADE environment can evolve dynamically, since agents can appear and disappear in the system according to the needs and the requirements of the application environment. Communication between peers, regardless of whether they are running in the wireless or wired network, is completely symmetric since each peer is able to play both the initiator and the responder role. JADE is fully developed in Java and it is based on the following driving principles [BEL 03]:

– *Interoperability*: JADE is compliant with the FIPA specifications [FIP 10]. As a result, JADE agents can interoperate with other agents, provided that they comply with the same standard.

– *Uniformity and portability*: JADE provides a homogeneous set of Application Programming Interfaces (APIs) that are independent of the underlying network and Java version. Actually, the JADE run-time provides the same APIs for the J2EE, J2SE and J2ME environment. In principle, it is possible for application developers to decide the Java run-time environment at deployment time.

– *Easy to use*: The complexity of the middleware is hidden behind a simple and intuitive set of APIs.

– *Pay-as-you-go philosophy*: Programmers do not need to use all the features provided by the middleware. Features that are not used do not require programmers to know anything about them, and do not add any computational overhead.

JADE offers the following list of features to the agent programmer:

– FIPA-compliant agent platform, which includes the agent management system (AMS), the default directory facilitator (DF) and the agent communication channel (ACC).

– Distributed agent platform. The agent platform can be split over several hosts. Only one Java application, and therefore only one Java virtual machine, is executed on each host. Agents are implemented as one Java thread and Java events are used for effective and lightweight communication between agents on the same host. Moreover, parallel tasks can be executed by one agent, and JADE schedules these tasks in a cooperative way.

– A number of FIPA-compliant additional DFs can be started at run-time in order to build multi-domain environments, where a domain is a logical set of agents, whose services are advertised through a common facilitator.

– Java API to send/receive messages to/from other agents; ACL messages are represented as ordinary Java objects.

– Lightweight transport of ACL messages inside the same agent platform, as messages are transferred encoded as Java objects, rather than strings, in order to avoid marshaling and unmarshaling procedures.

– Library to manage user-defined ontology and content languages.

– Graphical user interface (GUI) to manage several agents and agent platforms from the same agent. The activity of each platform can be monitored and logged. All lifecycle operations on agents can be performed through this administrative GUI.

4.3.3.1. *JADE architecture*

JADE includes both the libraries required to develop application agents and the run-time environment that provides the basic services that must be active on the device before agents can be executed. Each instance of the JADE run-time is called container (since it "contains" agents). The set of all containers is called platform (Figure 4.7) and provides a homogeneous layer, which hides the complexity and the diversity of the underlying layers (hardware, operating systems, types of network, and java virtual machine (JVM)) from agents and also from application developers [BEL 03].

4.3.3.2. *Behaviors to build complex agents*

The developer implementing an agent must extend JADE Agent class and implement agent specific tasks by writing one or more Behavior subclasses. User-defined agents inherit from their superclass the capability of registering and deregistering with their platform and a basic set of methods (e.g. send and receive agent communication language messages, use standard interaction protocols and register with several domains). Moreover, user agents inherit from their Agent superclass two methods: addBehavior(Behavior) and removeBehavior(Behavior), to manage the behavior list of the agent [BEL 01].

JADE contains established behaviors for the most common tasks in agent programming, such as sending and receiving messages and structuring complex tasks as aggregations of simpler ones. For example, JADE offers a so-called JessBehavior that allows full integration with JESS [JES 08], a scripting environment for rule programming offering an engine using the Rete algorithm to process rules.

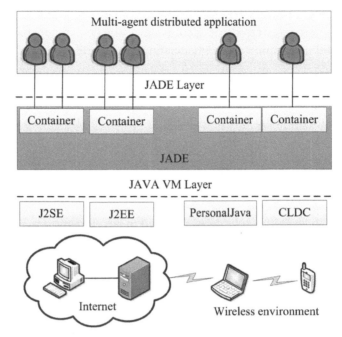

Figure 4.7. *JADE architectural model*

4.3.3.3. *JADE tools for platform management and monitoring*

Beyond a run-time library, JADE offers tools to manage the running agent platform and to monitor and debug agent societies; all these tools are implemented as FIPA agents themselves, and they require no special support to perform their tasks [BEL 01]. Examples of JADE tools are as follows:

– *Remote monitoring agent (RMA)*: the general management console for a JADE agent platform. The RMA acquires the information about the platform and executes the GUI commands to modify the status of the platform.

– *Directory facilitator (DF) GUI*: the DF agent also has a GUI, with which it can be administered, configuring its advertised agents and services.

– *The dummy agent* is a simple tool for inspecting message exchanges among agents, facilitating validation of agent message exchange patterns and interactive testing of an agent.

– *The sniffer agent* allows us to track messages exchanged in a JADE agent platform: every message directed to or coming from a chosen agent or group is tracked and displayed in the sniffer window, using a notation similar to Unified Modeling Language (UML) sequence diagrams.

The JADE is an interesting option that offers interoperability through the FIPA specification. It offers a friendly and robust support for the development of MASs that fulfill the requirements of the Horizon Project.

4.4. Context-aware technology for network control

Current Internet architecture has a unique protocol stack, the Transmission Control Protocol/Internet Protocol (TCP/IP) stack, running over the physical substrate. Thus, packets from all applications, each one with different performance requirements, are forwarded according to a single model with no differentiation. To provide customized services for the different applications in the future Internet, Horizon Project proposes a pluralist architecture, which allows different protocol stacks running simultaneously over the same shared physical substrate. The proposed architecture is based on the concept of network virtualization [AND 05]. The idea is to run multiple virtual networks on the same substrate providing resources for each network [HE 08] and, thus, we are able to create different networks, according to the requirements of applications running on each virtual network.

For the proposed architecture, the efficient sharing of the underlying network resources is a fundamental challenge. In the Horizon Project, we develop the infrastructure required for a piloting system to deal with the resource-allocation problem among the different virtual networks. The goal is to project the required information from the infrastructure onto the piloting system. Based on such information, the piloting system acquires the knowledge needed to optimize the creation and destruction of virtual networks and the distribution of the physical resources. For each network element, such as routers and switches, the piloting system defines a situated view that is used to determine the context surrounding the element and to select and optimize their control algorithms and parameters. The main idea is to define local algorithms, such as routing and QoS mechanisms, as a function of the context in order to increase network scalability. Hence, the piloting system must be aware of the context.

Context-aware systems are defined as systems that can be aware of the situation of entities or their contexts and act for them [LOK 06]. According to this definition, context and situation are closely related concepts and both are fundamental to understanding the context aware systems. We assume that context is "any information that can be used to characterize the situation of an entity", as defined by Loke [LOK 06]. Location, time, computational resources and network bandwidth are examples of context information. In addition, we assume that situation means a description of the current states of a given entity, as also defined by Loke [LOK 06]. For example, a link with slow responsiveness, high packet loss rate and high delay (context information) is probably congested (situation). In most of cases, as occurs in the previous example, we have to aggregate context information in order to determine the situation of an entity.

Context-aware systems must be able to determine the situations, as defined above, in which an entity is involved and to detect changes of situation being, as a consequence, aware of the context. Context awareness also enables the system to act automatically, which allows network control without human interference. To provide these functionalities, first, the system has to monitor the environment to acquire context information regarding a network element. After that, it has to reason about the context acquired and then it has to act in order to achieve a goal. Entities have also to communicate with other entities to determine their situations in time. To tackle this challenge, we propose to use the paradigm of an MAS as a modeling foundation [SIL 07]. The multi-agent paradigm seems to be attractive to develop an automatic piloting system because of the properties of agents themselves. These properties include autonomy, proactivity, adaptability, cooperating and mobility. In addition, multi-agents systems are decentralized in nature, which is required by large-scale networking environments. In this project, however, we prove the concept of such an automatic piloting system by evaluating a specific problem. It becomes reasonable to experiment the proposed solution in a representative case before expanding it to the large-scale problem because of the network complexity and the multitude of parameters the system must deal with. We aim at showing that it is possible to extrapolate our solution to the whole network evaluating its performance in a simpler case.

There are several techniques that can be used to develop context aware systems. This chapter describes some of these techniques and presents our choices for developing the piloting system. Section 4.4.1 describes a general layered architecture for context aware systems. The following sections present our choices based on the layers of this architecture. First, section 4.4.2 focuses on the context information that can be acquired from physical and virtual networks. After that, section 4.4.3 describes several techniques used to represent knowledge. Finally, section 4.4.4 defines the actions that can be taken to pilot the virtual networks. Concluding remarks are presented in section 4.5.

4.4.1. *Context-aware system architecture*

Context-aware systems have three basic functionalities: sensing, thinking, and acting. We follow the general abstract layered architecture proposed by Baldauf *et al.* [BAL 07] shown in Figure 4.8. In this general architecture, the three functionalities are represented by subsystems. Each subsystem is composed of one or more layers associated with each other in order to exchange information. Each subsystem, however, may be decoupled or tightly integrated into one device, that is a subsystem can think and act but uses a shared set of sensors to acquire context information. In addition, the levels of complexity of each one of these functionalities are independent. We can develop, for example, a system that has complex sensors but performs little reasoning before taking actions. Furthermore, the three basic subsystems can be implemented in a centralized or in a distributed fashion.

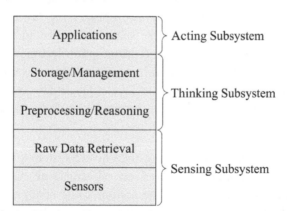

Figure 4.8. *A layered architecture model for context aware systems*

Sensing is a context aware subsystem that can be divided into two layers (Figure 4.8): sensors and raw data retrieval. The first layer consists of a set of sensors. We assume that sensors are every data source that provides context information no matter "what" the data source is. In this sense, a data source for temperature can be a hardware device, such as a thermometer, or a software module, such as a software application that requests the temperature to a Web service. Both are considered sensors because they provide temperature readings to the system. The second layer of the sensing subsystem is responsible for the retrieval of raw context data. This layer provides more abstract methods to the upper layers to request context information acquired by sensors. Actually, this layer is an interface to make implementation details of sensors transparent to the thinking subsystem. Hence, we can modify the sensors used by the system without modifications in the upper layers.

After collecting data with sensors, the task of the system is to use such data and to make sense of it [LOK 06]. This is the role of the thinking subsystem that can be divided in two layers: preprocessing and storage/management. The preprocessing layer converts all context information acquired into a common representation because the context information can be acquired in different forms, such as discrete values or a continuous series of values. Furthermore, this layer reasons and interprets context information to infer more knowledge. The reasoning technique used by the system ranges from simple event-condition rules to complex artificial intelligence techniques [LOK 06]. The preprocessing layer also aggregates context information from different sensors to provide more accurate information. In addition, knowledge representation techniques are needed to define and store context information in a machine processable form. This is the role of the storage/management layer. There are several knowledge representation techniques, such as graphical-based, logical-based and ontology-based techniques [BAL 07].

The third subsystem is acting. A context aware system takes actions based on the context information gathered or on situations recognized by the sensing and thinking subsystems. The actions to be taken by the system are defined according to the application requirements, as shown in Figure 4.8. In a congestion control application, for example, sensors monitor the bandwidth in a network link and indicate that the link is saturated. Thus, the system "thinks" and decides to block new flows and to maintain all the current flows. In general, a context aware system must act in time to quickly adapt its operation according to environmental changes. In addition, these systems allow users to control actions, i.e. users should be able to override, cancel and stop actions and also reverse the results of an action.

We follow the described architecture to develop a network piloting system. We implement the layers in a distributed fashion based on MASs. The sensing subsystem is implemented by sensors spread along the virtual routers. Thus, we do not have a sensor per agent but a set of sensors that can be used by all the agents. Sensors collect context information by directly reading data available on the operational systems of the routers and also by using measurement tools. We have also defined that sensors translate the acquired context information into a common representation. The context information considered and the sensors under development are detailed in section 4.4.2. The agents implement all the other layers related to thinking and acting subsystems. Thus, they must communicate with the set of sensors in order to acquire context information and, thus, process this information to take actions. We discuss knowledge representation techniques in section 4.4.3 and also the actions defined by the piloting system in section 4.4.4.

4.4.2. *Sensing subsystem*

We develop a set of sensors decoupled of the agents. Thus, we are able to develop lightweight agents in terms of computational processing because they do not have to implement a sensor per context information to be sensed. The agents only have to exchange messages with sensors to request and to receive the context information desired. Therefore, although we slightly modify the classical architecture of agents by decoupling the sensing subsystem, every agent is still capable of acquiring sensing information.

The sensors that we are developing basically acquire context information by using two methods. Sensors read available data provided by operational systems running on physical and virtual routers and also use well-known tools for monitoring networks, such as ping [LIN 10], nmap [NMA 10], and ifconfig [VAN 10]. After collecting data, sensors translate the raw data into Extended Markup Language (XML) [QUI 10] data structures that are ready to be sent to agents when requested. In this chapter, we focus on the description of the context information that is useful for controlling the resource allocation among virtual networks. To virtualize a network, we first define the resources we are planning to share among the multiple virtual networks. Currently, we have identified the following basic resources: processing power, memory, bandwidth, traffic and network topology [SHE 09]. The computational resources of the network elements, such as routers and switches, must be sliced among the virtual networks. These resources include processing power and memory. It is worth mentioning that the resource sharing is a fundamental requirement to make the system work properly. A router can stop to forward packets and to exchange routing messages if it has no CPU cycle available. Memory isolation is an important requirement to avoid the overriding of one routing table on another virtual router when a routing table of a given virtual router increases and, thus, packets will be not correctly forwarded. Bandwidth, traffic and topology are the network resources that we have to slice. First, we have to guarantee to each virtual network its own fraction of bandwidth on a physical link. All networks elements on a given source–destination path that are able to limit the forwarding rate should also be sliced [SHE 09]. Second, we have to be able to associate a specific set of traffic with one or more virtual networks, i.e. a set of traffic must be isolated from another. This is a key point to employ in multiple virtual networks. In this sense, traffic means all packets to/from a given address or all HyperText Transfer Protocol (HTTP) traffic, for example. Finally, each network element should be aware of the nodes within a given virtual network. Thus, a virtual router should be able to determine its own view of other routers and the connectivity among them.

We developed sensors to acquire context information related to the basic resources. Thus, the data that we can acquire from physical and virtual networks are divided in two groups: computational resources and network states. The first group is related to each physical and virtual router. On the other hand, the second group indicates the states of each physical and virtual link. For each physical router, our sensors are currently able to acquire the following data depending on the network virtualization technique employed:

– Processor usage (physical CPUs).

– Used memory.

– Available memory.

– Used swap memory.

– Available swap memory.

– Total memory allocated to a given virtual router.

– Number of virtual routers per physical router.

– Number of virtual processors allocated to a given virtual router.

– Number of virtual interfaces defined to a given virtual router.

Sensors directly read most of the required data from the operational system of physical routers. We estimate physical processor usage by summing the usage of virtual processors.

For each virtual router, the following data are acquired by sensors:

– Processor usage (virtual CPUs).

– Used memory.

– Available memory.

– Used swap memory.

– Available swap memory.

According to the employed virtualization technique, the virtual CPU usage is obtained by using tools provided by the operational system of physical routers; and the memory information is directly read from the operational system of virtual routers. In this case, we need to access these virtual routers.

The second group of context information is related to network states. Our sensors are currently able to acquire the following information per physical or virtual network interface:

– Number of received packets.

– Received bytes.

– Number of erroneous packets.

– Number of dropped packets during receptions.

– Receiving rate.

– Number of transmitted packets.

– Transmitted bytes.

– Number of dropped packets during transmissions.

– Transmission rate.

Sensors also discover the topologies of physical and virtual networks. Thus, for each neighbor, we can determine, for example, the available bandwidth and the latency of a given link. In our project, we consider that the context information previously presented is enough to serve as a basis for the reasoning techniques employed by the piloting system under development. With this information, the piloting system will be able to react to environmental changes in order to guarantee the isolation and the resources of each virtual network.

4.4.3. *Thinking subsystem*

The role of the thinking subsystem is to make sense of data acquired by sensors and to use context information to react to environmental changes. Basically, thinking subsystems combine knowledge representation and reasoning techniques. Knowledge representation techniques define and store context information in a machine processable form [DAV 93]. Thus, the goal of a knowledge representation technique is to store the context information in a logical form in order to allow reasoning techniques to use this information. The reasoning techniques include, for example, mathematical models, inference techniques and cognitive-based models [LOK 06]. We are focusing on the knowledge representation techniques. Several techniques are proposed to represent knowledge and there are studies that classify these techniques into classes [LOK 06, STR 04, BAL 07]. Four of the most relevant classes of knowledge representation techniques are briefly described in the following paragraphs.

Markup-based techniques define hierarchical data structures to represent context information. These data structures are composed of markup tags with attributes and content. The content of each tag might be recursively defined by other tags. One of the most popular markup languages is XML [QUI 10]. Currently, there are several XML-based knowledge representation languages and standards, such as DARPA Agent Markup Language (DAML) [DAR 10] and Web Ontology Language (OWL) [MCG 10]. We are currently using XML as the common language to represent all data acquired by sensors and also to describe the content of the exchanged messages in the system.

In general, a logic derives a concluding expression from a set of expressions or facts based on predefined conditions [STR 04]. This process is called inference and the conditions are formally described by a set of rules. Consequently, context information is represented by facts, expressions and rules if we are considering logic-based techniques. Thus, context information is added to a logic-based system in terms of facts as well as being deleted or updated. Context information may be also inferred from the rules defined in the system. Ranganathan and Campbell [RAN 03], for example, propose to represent context and situations by using first-order logic techniques. In this case, rules are defined to map situations onto actions by using Prolog language. The proposed rules basically relate context information to situations. Context information is the condition of a rule and situations are the conclusions of a rule.

Graphical modeling techniques are largely used because of their intuitive nature. The UML [OBJ 10] is a well-known general-purpose modeling tool and can also be used to represent context information [KOG 02]. The UML class diagrams are the graphical components of this language and, from these diagrams, we can derive entity-relationship models [BER 05]. This kind of model is largely used as structuring tool for developing relational databases, which can be viewed as a knowledge base [STR 04]. Another graphical technique is called contextual graphs [BRÉ 03, PAD 04]. This technique does not define diagrams, like UML does, but it proposes the concept of context spaces, which is a spatial view of context information. Each type of context information is represented by one axis of a multidimensional space and, thus, sensors' readings are represented by points and situations are represented by regions in this space.

Ontology is generally defined as a set of concepts and terms that are used to describe a domain of knowledge or to develop a representation of this domain, including the relationships between their elements. Particularly, the set of terms can be ordered in a hierarchical fashion and used as a "sketch" to build a knowledge base [GÓM 99]. From this definition, we clearly identify the differences between ontologies and knowledge bases. Ontology provides a set of terms to describe a given domain while a knowledge base uses these terms to describe a given situation. If this situation changes, the knowledge base also changes. The ontology, however, does not change because the domain remains the same. Thus, we can easily develop a knowledge base from ontology, which is the main advantage of this technique. Furthermore, the ontology-based techniques have other advantages. First, these techniques avoid uncertainty because they provide an exact description and a specific vocabulary to represent knowledge. In addition, ontology-based techniques allow knowledge sharing because applications within the same domain can use the same ontology. Finally, the same ontology can be expressed in different languages. In general, an ontology is composed of a taxonomy, i.e. a set of concepts and a hierarchy between them; a set of relationships between these concepts and a set of axioms [GÓM 99].

We built a knowledge base for the piloting system, thus we define an ontology to describe the environment with multiple virtual networks. In addition, most of the analyzed platforms for developing agents have tools to define ontologies inside their agents. The Ginkgo platform [GIN 08, GIN 09] considers an ontology-based representation composed of *classes* and *individuals*. Both concepts are quite similar to the classes and instances of an object data-model. An individual is an instance of a class. Hence, a class is a set of individuals, which are members of this class, and the knowledge base is a tree of classes, as shown in Figure 4.9 (adapted from [GIN 09]). In this example, we have two classes, *person* and *employee*, that derive from the root class, *thing*, and one instance of each class, *John* and *Jane*. In practice, classes can be routers, users, flows, applications, etc. Furthermore, we can observe that classes have properties to store data. These properties have an identifier and are single or multivalued. The individuals have the same properties of the class of which they are members and also the properties of their parent calsses.

4.4.4. *Acting subsystem*

The acting subsystem is the part of the piloting system responsible for reacting to environmental changes based on the context information gathered or on situations recognized by the sensing and thinking subsystems. Our choice is to define typical basic functionalities regardless the virtualization technique employed by the network. The actions taken by the piloting system are then based on these functionalities, which are described in the following.

We identified four basic functionalities: creation of multiple customized networks, flexible management, real-time control and monitoring. We also defined primitives in order to employ these functionalities. The primitives allow to *instantiate/delete* and also to *migrate* network elements and flows and *set* their resource-allocation parameters. Such primitives make the network virtualization a suitable technology for creating multiple virtual networks and, as a result, for supporting the pluralist approach, because several requirements are satisfied, as explained below.

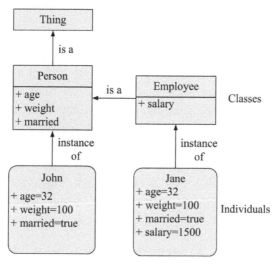

Figure 4.9. *A simple knowledge base using the Ginkgo platform*

In a pluralist architecture, we have multiple customized networks running in parallel and, thus, the creation functionality is fundamental. In this sense, the instantiate primitive can be used to instantiate virtual network elements, such as virtual routers and/or virtual links, and, therefore, multiple virtual networks can be rapidly deployed and run simultaneously. Each virtual network has its own protocol stack, network topology, administration policy, etc. This enables network innovation and new business models [FEA 07]. With network virtualization, a service provider can allocate an end-to-end virtual path and instantiate a virtual network tailored to the offered network service, for example a network with QoS guarantees. Hence, new services can be easily deployed and new players can break the barrier to enter into the network service market.

Flexible management is the second functionality that we have identified. The network virtualization layer breaks the coupling between the logic used to construct the forwarding table and the physical hardware that implements the packet forwarding task [WAN 08]. Therefore, the migrate primitive allows moving a virtual network element from one physical hardware to another, without changing the logical/virtual network topology. In addition, traffic engineering and optimization techniques can use the migrate primitive to move virtual network elements/links along the physical infrastructure in order to minimize energy costs, distance from servers to specific network users or other objective functions.

Real-time control is the third functionality. The virtual networks architecture also supports real time control of virtual network resources because resource-allocation parameters can be set for each virtual network element (router, switch, link, gateway, etc.). We can set the allocated memory, bandwidth, maximum tolerated delay, etc. Even specific hardware parameters can be set, for instance the number of virtual processors and priority of processor usage in a contention scenario. Thus, we can dynamically adapt the resources allocated to each virtual network according to the current network condition, number of users, priority of each virtual network, service level agreements (SLAs), etc.

Monitoring is also an important functionality because network virtualization techniques require a set of monitoring tools required to measure variables of interest, such as available bandwidth, processor and memory usage and link and end-to-end delay. The monitor primitive is called to measure the desired variables. In this project, the measurements are performed by sensors as explained in section 4.4.2. Thus, when the monitor primitive is called actually the sensors are called.

The four functionalities can be used by the piloting system to guarantee the requirements of each virtual network. For example, a given virtual network employs an intrusion detection system (IDS) in order to detect malicious nodes. In this case, the delete primitive can be used to delete a virtual network element/link or even an entire network if an attack (e.g. distributed denial of service (DDoS)) is detected. Figure 4.10 also shows an example of traffic engineering that uses the migrate primitive to move virtual routers along the physical infrastructure to minimize energy costs or other objective functions. In this example, we have five physical routers (R_1, R_2, R_3, R_4 and R_5) and initially two virtual networks. The first virtual network provides QoS for voice-over-IP (VoIP) calls and is composed of virtual routers A_1, A_2 and A_3 placed, respectively, in physical routers R_1, R_2 and R_3. The second virtual network is a secure network composed of virtual routers S_1, S_2 and S_3 located, respectively, in physical routers R_4, R_2 and R_5. Assume that a load balance rule moves the virtual router with the highest

CPU usage just after the physical CPU usage reaches 80%. In this scenario, shown in Figure 4.10a), the physical router R_2 has two virtual routers A_2 and S_2 and each router uses 25% of the physical CPU. Thus, the defined rule is satisfied. Suppose that a third virtual network is created to provide QoS for video streaming applications. As shown in Figure 4.10b), this network is composed of virtual routers V_1, V_2 and V_3 located, respectively, in physical routers R_4, R_2 and R_5. Thus, we add one virtual router to R_2 and V_2 requires more than 25% of the physical CPU usage. Currently, the CPU usage is 75% and the rule is still satisfied. After that, however, the virtual router V_2 requires more than 10% of CPU usage and the total CPU usage becomes greater than 80%. Thus, according to the rule defined, we have to move the virtual with the highest CPU usage, which is V_2 in this case. An alternative is to move V_2 to R_3, as shown Figure 4.10c). After migrating the virtual router, the CPU usage in all physical routers is less than 80% and the rule is satisfied.

4.5. Summary

This chapter describes the state of the art on autonomic systems as well as MASs. These systems are well adapted to complex environments and are composed of highly dynamic networks that implement the concept of self-piloting systems. We give an overview of three platforms for building agents that can be useful for the development of network control and management.

We develop the infrastructure required for a piloting system to control the resources allocated to each virtual network. This system creates and destroys virtual networks, sets their parameters and migrates network elements. These actions are executed based on the context information acquired by sensors spread along network elements. The piloting system is based on the multi-agent paradigm, developed in a distributed fashion to increase network scalability.

The piloting system also follows the layered architecture discussed in section 4.4.1. The sensors under development read available data provided by operational systems running on physical and virtual routers and also use well-known tools for monitoring networks. Sensors, however, are decoupled from agents to make them lightweight in terms of computational processing. In this case, agents only have to exchange messages with sensors to request and to receive the context information desired. Currently, sensors do not send messages to agents. After collecting data, sensors translate the raw data into XML data structures that are sent to a server.

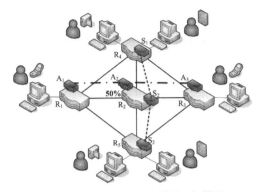

a) Two virtual networks and 50% of CPU usage in R_2

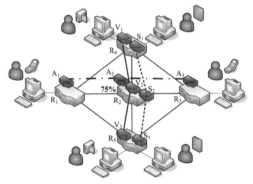

b) Three virtual networks and 75% of CPU usage in R_2

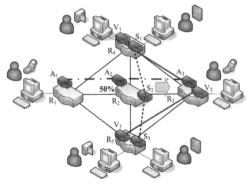

c) Virtual router V_2 migrates to physical router R_3

Figure 4.10. *Traffic engineering using the migrate primitive*

The actions defined for the piloting system are the creation of multiple customized networks, flexible management, real-time control and monitoring. In addition, the primitives defined to employ these functionalities are instantiate, delete and migrate network elements and flows, and set the resource-allocation parameters. These functionalities and primitives are feasible regardless of the virtualization technique employed by the network. An experimental analysis is required to evaluate if these functionalities and the defined network resources are enough to guide the network.

4.6. Acknowledgments

We would like to thank Carlos Roberto Senna, Daniel Macêdo Batista and Nelson Luis Saldanha da Fonseca for their work in improving the final version of this chapter.

4.7. Bibliography

[AND 05] ANDERSON T., PETERSON L., SHENKER S., et al., "Overcoming the internet impasse through virtualization", *IEEE Computer*, vol. 38, pp. 34–41, April 2005.

[BAL 07] BALDAUF M., DUSTDAR S., ROSENBERG F., "A survey on context-aware systems", *International Journal of Ad Hoc and Ubiquitous Computing*, vol. 2, pp. 263–277, June 2007.

[BEL 01] BELLIFEMINE F., POGGI A., RIMASSA G., "Developing multi-agent systems with JADE", *Intelligent Agents VII Agent Theories Architectures and Languages*, vol. 1986, pp. 89–103, 2001.

[BEL 03] BELLIFEMINE F., CAIRE G., POGGI A., et al., "JADE a white paper", *Exp*, vol. 3, no. 3, pp. 6–19, September 2003. Available at http://jade.cselt.it/papers/2003/WhitePaperJADEEXP.pdf (accessed in May 2013).

[BER 05] BERARDI D., CALVANESE D., DE GIACOMO G., "Reasoning on UML class diagrams", *Artificial Intelligence*, vol. 168, nos. 1–2, pp. 70–118, 2005.

[BRÉ 03] BRÉZILLON P., "Representation of procedures and practices in contextual graphs", *The Knowledge Engineering Review*, vol. 18, pp. 147–174, June 2003.

[BRI 01] BRIOT J., DEMAZEAU Y. (eds), *Principes et Architecture des Systèmes Multi-Agents*, Hermes, 2001.

[BOR 05] BORDINI R.H., DASTANI M., DIX J., et al., (eds), *Multi-Agent Programming Languages, Platforms and Applications*, Springer, 2005.

[BUL 08] BULLOT T., KHATOUN R., HUGUES L., et al., "A situatedness-based knowledge plane for autonomic networking", *International Journal of Network Management*, vol. 18, pp. 171–193, March 2008.

[DAR 10] DARPA TEAM, *The DARPA Agent Markup Language Homepage*, June 2010. Available at http://www.daml.org (accessed in May 2013).

[DAV 93] DAVIS R., SHROBE H., SZOLOVITS P., "What is a knowledge representation?", *AI Magazine*, vol. 14, no. 1, pp. 17–33, 1993.

[DOB 06] DOBSON S., DENAZIS S., FERNÁNDEZ A., *et al.*, "A survey of autonomic communications", *ACM Transactions on Autonomous and Adaptive Systems*, vol. 1, pp. 223–259, December 2006.

[FAC 06] FACI N., GUESSOUM Z., MARIN O., "DimaX: a fault-tolerant multi-agent platform", *Proceedings of the 2006 International Workshop on Software Engineering for Large-Scale Multi-Agent Systems, SELMAS '06*, ACM, New York, NY, pp. 13–20, 2006.

[FEA 07] FEAMSTER N., GAO L., REXFORD J., "How to lease the Internet in your spare time", *ACM SIGCOMM Computer Communication Review*, vol. 37, pp. 61–64, January 2007.

[FER 95] FERBER J., *Les Systèmes Multi-Agents: Vers Une Intelligence Collective*, Dunod, 1995.

[FIP 10] FIPA, "The foundation for intelligent physical agents", October 2010. Available at http://www.fipa.org (accessed in May 2013).

[GIN 08] Ginkgo Networks, Ginkgo distributed network piloting system, Technical report, Ginkgo Networks, September 2008.

[GIN 08] Ginkgo Networks, "White paper–Ginkgo distributed network piloting system", September 2008. Available at http://www.ginkgo-networks.com/ IMG/pdf/WP_Ginkgo_DNPS_v1_1.pdf (accessed in May 2013).

[GIN 09] Ginkgo Networks, Ginkgo agent platform programming manual v1.5, Technical report, Ginkgo Networks, October 2009.

[GÓM 99] GÓMEZ-PÉREZ A., "Ontological engineering: a state of the art", *Expert Update*, vol. 2, no. 3, pp. 33–43, 1999.

[HE 08] HE J., ZHANG-SHEN R., LI Y., *et al.*, "DaVinci: dynamically adaptive virtual networks for a customized internet", *Proceedings of the ACM CoNEXT Conference*, ACM, Madrid, Spain, December 2008.

[IBM 06] IBM, An architectural blueprint for autonomic computing, 4th ed., White Paper, June 2006.

[JES 08] JESS, "Jess, the rule engine for the java platform", November 2008. Available at http://herzberg.ca.sandia.gov/jess/ (accessed in May 2013).

[KOG 02] KOGUT P., CRANEFIELD S., HART L., *et al.*, "UML for ontology development", *The Knowledge Engineering Review*, vol. 17, no. 1, pp. 61–64, 2002.

[LIN 10] Linux.org, *Ping Man page*, June 2010. Available at http://linux.die.net/man/8/ping (accessed in May 2013).

[LOK 06] LOKE S., *Context-Aware Pervasive Systems: Architectures for a New Breed of Applications*, 1st ed., Auerbach Publications, 2006.

[MCG 10] McGUINNESS D.L., VAN HARMELEN F., *OWL Web Ontology Language Overview*, June 2010. Available at http://www.w3.org/TR/owlfeatures/ (accessed in May 2013).

[NMA 10] Nmap.org, *Nmap Reference Guide*, June 2010. Available at http://nmap.org/book/man.html (accessed in May 2013).

[OBJ 10] OBJECT MANAGEMENT GROUP, *UML Resource Page*, June 2010, Available at http://www.uml.org/ (accessed in May 2013).

[PAD 04] PADOVITZ A., LOKE S.W., ZASLAVSKY A.B., "Towards a theory of context spaces", *Proceedings of the Workshop on Context Modelling and Reasoning (COMOREA)*, pp. 38–42, IEEE, March 2004.

[QUI 10] QUIN L., *Extensible Markup Language (XML)*, June 2010. Available at http://www.w3.org/XML/ (accessed in May 2013).

[RAN 03] RANGANATHAN A., CAMPBELL R.H., "An infrastructure for contextawareness based on first order logic", *Personal and Ubiquitous Computing Journal*, vol. 7, pp. 353–364, December 2003.

[SHE 09] SHERWOOD R., GIBBY G., YAPY K.-K., *et al.*, FlowVisor: a network virtualization layer, Technical report, Deutsche Telekom Inc. R& D Lab, Stanford University, and Nicira Networks, October 2009.

[SIL 07] SILVA V., LUCENA C.J.F., "Modeling multi-agent system", *Communications of the ACM*, vol. 50, pp. 103–108, May 2007.

[STE 03] STERRITT R., BUSTARD D., "Towards an autonomic computing environment", *Proceedings of the 14th International Workshop on Database and Expert Systems Applications, DEXA '03*, IEEE Computer Society, p. 699, 2003.

[STR 04] STRANG T., LINNHOFF-POPIEN C., "A context modeling survey", *Proceedings of the International Workshop on Advanced Context Modelling, Reasoning and Management*, The 6th International Conference on Ubiquitous Computing (UbiComp 2004), Nottingham, England, 2004.

[VAN 10] VAN KEMPEN F.N., COX A., BLUNDELL P., *et al.*, *Ifconfig Man page*, June 2010. Available at http://linux.die.net/man/8/ifconfig (accessed in May 2013).

[WAN 08] WANG Y., KELLER E., BISKEBORN B., *et al.*, "Virtual routers on the move: live router migration as a networkmanagement primitive", *ACM SIGCOMM Computer Communication Review*, vol. 38, no. 4, pp. 231–242, 2008.

[WOO 02] WOOLDRIDGE M., *An Introduction to Multi-Agent Systems*, John Wiley & Sons, 2002.

Chapter 5

Providing Isolation and Quality-of-Service to Virtual Networks

This chapter presents a proposal for handling virtual network requirements, such as isolation, Quality-of-Service (QoS) and congestion and failures avoidance. Data networks represent a dynamic and complex area, in which managers face new problems and challenges every day. The rapid growth of network applications and the increasing amount of information collected make resources and network control more and more complex. Conflicting objectives between infrastructure providers and virtual networks, and between coexisting virtual networks sharing the same physical substrate and also changing requirements make the management and control of virtual network a difficult task. Therefore, in such an unpredictable, changing and open environment, a dynamic control can be a solution to achieving an improved network management and monitoring. Adaptive network monitoring approaches are then proposed to monitor the network state and orchestrate its different components. The main goal is to add an intelligent control to the network to guarantee QoS and, at the same time, improve network management and overall performance.

In this chapter, section 5.1 describes existing control algorithms, which can be used in virtual networking. Section 5.2 describes the main challenges for

---

Chapter written by Miguel Elias M. CAMPISTA.

packet forwarding using Xen. Then, a proposal developed for local control within physical network nodes is presented in section 5.3. This proposal is designed within the piloting plane to guarantee that agreements from all virtual networks are respected, even in the presence of misbehaving virtual networks that violate their service level agreements (SLAs). The main idea is to control domain 0 (dom0) shared resources, adjusting the amount of resources used per virtual network according to their SLAs [FER 11]. Finally, we provide final remarks in section 5.4.

5.1. Background on control and management of virtual networks

Virtual networking control and management can be handled in local or in global scope. In global scope, operations such as instantiating virtual nodes and virtual links, and migrating virtual networks are included. In local scope, on the other hand, resources assigned to each virtual network are monitored on each physical node. For instance, the local control must deal with virtual network isolation.

Schaffrath *et al.* propose an architecture for global control of virtualized networks [SCH 09]. The proposal, implemented using Xen, assumes a centralized control that creates network slices through the instantiation of virtual machines and virtual links. Thus, on receiving a request to allocate a new virtual network, the system contacts each of the selected physical nodes and instantiates the virtual machine needed to build the entire network as well as the virtual links between them using Internet Protocol (IP) tunneling. Other similar approaches to global control are found in virtualization-based testbeds, such as GENI [GEN 08]. The access of the researcher to the testbeds is controlled through a central entity called the clearing house. The clearing house monitors the physical nodes and the services available on each one of the federated testbeds, who is authorized to use them and which slices are scheduled for each researcher [GUI 13].

Entities with global control also perform other functions, such as migration [CAR 12]. Houid *et al.* propose a global system based on multi-agents to dynamically allocate resources to virtual networks using migration [HOU 10]. The system monitors the available resources on each physical machine as well as the changing requirements of the virtual networks. As resources are scarce, the agent on the physical node searches a similar physical node to receive one or more of its virtual machines.

Global network control systems do not deal with resource sharing within a physical machine, instead they assume that slices are isolated by a local control mechanism. Egi *et al.* [EGI 07] investigate the construction of a platform composed of virtual routers using Xen and Click [KOH 99]. They evaluate the provision of isolation and fairness among the networks using local control. Authors investigate the use of different data planes, assuming routing through a privileged domain and through virtual machines, and evaluate the ability to share resources among virtual networks. Egi *et al.* extend the CPU scheduler of Click to evaluate the CPU cost of forwarding a packet. Their work, however, does not define management capabilities to specify the amount of resources for each virtual network. Also, they do not tackle traffic differentiation to ensure the QoS in virtual networking.

An important aspect of local control is the guarantee of isolation among virtualized environments. Xen has problems, not only concerning input/output (I/O) operations but also in other aspects such as fairness [HAN 09, JIN 09]. Jin *et al.* propose a mechanism to ensure fairness on L2 and L3 caches' utilization in Xen, which is not considered by the isolation mechanisms of Xen hypervisor [JIN 09]. The proposed algorithm modifies the allocation of memory pages by the hypervisor, using the technique of page coloring.

McIlroy and Sventek propose a local control based on Xen for the current Internet. Each flow that requires QoS is allocated to a virtual machine, called QoS routelet [MCI 06]. Each virtual machine then applies its QoS policies across the incoming traffic. The prototype is implemented using Xen, and the traffic without QoS requirements is routed by the privileged domain, whereas traffic with QoS is routed by virtual machines. The authors note that it is not possible to guarantee QoS in the strictest sense with this model because the Xen scheduler is not suitable for this task. Another shortcoming of this proposal is scalability because it needs a virtual machine per QoS flow.

Mechanisms for isolating virtual environments have also been proposed for other virtualization platforms. Trellis [BHA 08] is a system to provide isolation on the virtual network infrastructure (VINI) [VIN 10], which is a testbed similar to PlanetLab. The difference is that VINI, unlike PlanetLab, acts as a private testbed within their project. Because VINI is based on virtualization on the operating system level, i.e. all virtual environments share the same kernel, the performance of packet forwarding using Trellis is lower than that of Xen with plane separation [EGI 08]. The major problems of these approaches [BHA 08, WAN 08, ZEC 03] are the execution of all control planes on the same operating system and the lack of support for creating differentiated data planes for each virtual network.

Genesis is a kernel to create virtual networks with different architectures [KOU 01]. Based on the concepts of hierarchy and inheritance, Genesis proposes that different "child" virtual networks should be created based on a "root" network, from which the children inherit common characteristics. Similar to Trellis, Genesis is based on the premise that all control planes work on the same operating system. It allows, however, the usage of different policies and QoS mechanisms for each virtual network. Because Genesis is implemented at the user level and inserts a virtualization layer, it provides low routing performance.

Another virtualization platform is OpenFlow [MCK 08], which is based on a network of simple forwarding elements and a centralized control plane. To share the physical resources of the forwarding elements among the virtual networks, OpenFlow provides the FlowVisor tool [SHE 10]. The FlowVisor can be seen as a transparent proxy between the forwarding elements and the control planes. FlowVisor controls the use of CPU and memory in the forwarding elements as well as the division of the network space, i.e. which characteristics define each virtual network.

Another approach based on Linux and Click to create a shared data plane is proposed by Keller and Green [KEL 08]. In this proposal, each virtual network can create its own data plane, based on generic assumptions for packet forwarding in Click. However, the authors do not address a fair division of resources among the control planes.

The proposal described in this chapter is focused on local and global control. The local control is performed by works that guarantee the isolation and the SLAs within each physical node in Xen networks. The global control autonomously solves physical node failures using a distributed fault-tolerant algorithm and also migrates virtual networks hosted on overwhelmed physical nodes.

5.2. Challenges in packet forwarding using Xen

The virtual network model using Xen considers that virtual machines behave as routers[1]. A virtual network is defined as a set of virtual routers and links, created over the physical infrastructure, as illustrated in Figure 5.1. The Xen architecture is composed of a hypervisor, virtual machines, called user domains (domUs), and a privileged virtual machine called dom0. The Xen hypervisor controls the access to the physical resources and handles the I/O operations performed by the domains. dom0 is a privileged domain that directly accesses the hardware. Because dom0 is a driver domain, it stores all physical device drivers and creates an interface between the virtual drivers placed in the domUs and the physical devices. In addition, dom0 is also the management interface between the administrator and the hypervisor to create virtual machines, change Xen parameters and manage Xen operation [PIS 11].

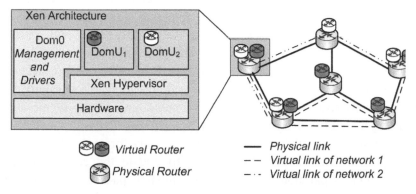

Figure 5.1. *Xen architecture with two virtual networks*

Sending and receiving packets are I/O operations, which require the use of the device drivers located at dom0. Thus, all network operations from domUs generate an overhead in terms of both memory and CPU of dom0. The Xen hypervisor, however, does not efficiently isolate dom0 resource utilization, which is a major vulnerability of Xen [MAT 12]. Table 5.1 shows that a domU can easily increase dom0 CPU consumption by performing network

---

1 A previous version of this work is available in [FER 11].

operations[2]. Since data transfer between two domUs and between a domU and the dom0 consumes dom0 CPU resources, a malicious or fault action in a domU can easily exhaust dom0 resources, compromising the performance of all the other domains. Hence, the goal of management and control in virtualized environments is to avoid operations conducted on a virtual network, which affect the others breaking down the isolation premise.

CPU (%)	Description
0.71 ± 0.60	Basic CPU consumption on $dom0$
66.43 ± 8.93	TCP traffic from $domU$ to $dom0$
85.49 ± 5.91	TCP traffic from $domU_1$ to $domU_2$
1.79 ± 1.01	TCP traffic from an external machine to $domU$

Table 5.1. *CPU consumption on dom0*

The Xen conventional architecture is not efficient for network operations because domU packet forwarding takes a long and slow path. As depicted in Figure 5.2a), the packet arrives at dom0, follows to domU and returns to dom0 to be forwarded to the next router. The plane separation paradigm is an alternative way to improve the forwarding performance because packets are directly forwarded by a shared data plane in dom0, as shown in Figure 5.2b). The plane separation is accomplished by maintaining a copy of the current forwarding table of domU in dom0, which has direct access to the hardware. It is important to note that data packets are directly forwarded by dom0, but control packets are forwarded to domU to update the control plane. In addition, plane separation does not avoid flexible packet forwarding. If a virtual router requires customized operations, they cannot be supported by the shared data plane, such as monitoring or modifying a specific header field. In this case, the plane separation can be ignored by inserting a default route to the virtual machine in the forwarding table within dom0, as done in conventional packet forwarding.

---

2 These results were measured with the `top` tool in a machine with Intel Core 2 Quad processor with 4 GB of RAM and Xen 3.4-amd64. Each domU is configured with one virtual CPU and 128 MB of memory, and dom0 is configured with one virtual CPU and no memory constraints. Each virtual CPU is associated with an exclusive physical CPU. TCP traffic was generated with `Iperf`. The basic CPU consumption indicates the CPU usage in dom0 when there are no operations in domUs. A confidence interval of 95% is assumed.

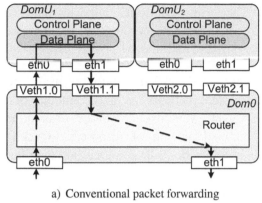

a) Conventional packet forwarding

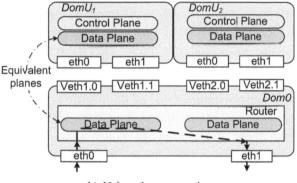

b) Using plane separation

Figure 5.2. *Packet forwarding in Xen-based networks*

Finally, Xen also does not provide any scheme for QoS provision. Thus, a scheme to control traffic policies must be built to guarantee QoS within a virtual network and also among them.

5.3. Controlling Domain 0 shared resources

Domain 0 resources, such as CPU, memory and bandwidth, are shared by the virtual routers in I/O operations. Because there is no tight control in this resource sharing, these resources may be depleted during packet forwarding, and isolation among virtual routers may be broken. The isolation failure may incur security problems as well as in SLA violations. Hence, it is necessary to develop mechanisms for local control to guarantee each virtual network SLA in each physical node. The main objective of the presented controller is to provide isolation to the Xen virtualization platform by controlling the use of dom0 resources. Then, this controller allocates and monitors physical resources used by all domUs according to the parameters set by the administrator of the physical machine.

The controller presented, called maximum usage controller (MUC), reserves, for each virtual network, a fixed amount of resources and a parameter called weight. This fixed resource reservation guarantees the availability of a minimal amount of resources to each virtual network, whereas the weight specifies how idle resources are distributed among virtual networks. As long as there are idle resources, these are provided to demanding networks. Indeed, this is the police chosen to control physical resource utilization. In addition, this controller is based on resource monitoring and punishment. If the punishment is applied, its severity is according to the level of violation of each virtual network SLA value. Also, it supports the plane separation paradigm. Hence, resources are correctly monitored, independent of the data plane location.

The following analysis assumes that all domUs are not trustworthy, because each domU has a different administrator. Hence, a domU may, intentionally or not, harm other domains. In the proposed model, a domU can be malicious or present an unacceptable behavior. A malicious behavior occurs when a domU intentionally executes an action to break the isolation among virtual networks and then damage other domUs' performance. An unacceptable behavior, on the other hand, is any attempt from a domU to exceed the amount of resources reserved to it that uses all dom0 idle resources. This behavior could interfere in the operation of other domains. Because both behaviors are harmful, virtual machines executing any one of them can be classified as opponent domains, whereas the others are classified as common domains.

5.3.1. *Maximum usage controller*

The MUC [FER 11] allocates and monitors the total usage of dom0 resources, $U(t)$, and the respective usage by each virtual router i, $U_i(t)$, in every T seconds. The dom0 resources can be allocated either by fixed reservation or on demand. The allocation based on fixed reservation is handled by the administrator, who reserves a fixed amount of dom0 resources for each domU, ensuring a minimum quality for each virtual network. The on-demand allocation, in opposition, aims at guaranteeing high efficiency in resource usage because MUC redistributes the idle resources among the domUs with a demand greater than the reserved fixed amount. Idle resources are composed of the non-reserved resources in addition to the reserved resources that are not in use. Thus, a premise of the controller is to provide the fixed resources of a virtual router i, denoted as a percentage α_i of the total dom0 resources, $R(t)$, whenever there is a demand. Another premise is to allocate all idle resources on demand to virtual routers according to the priority previously assigned by the administrator. This priority is a parameter called weight, denoted by W_i, where $\{W_i \in \mathbb{Z} \mid 1 \leq W_i \leq 1,000\}$. The higher the weight of a virtual router, the more idle resources on dom0 it has access to. Thus, the on-demand allocation provides an additional differentiated service for each virtual network.

MUC monitors bandwidth by observing the volume of bits transmitted on each output physical link. If a router exceeds the allocated bandwidth in an output link, it is punished by having its packets (sent on that link) dropped.

The CPU usage in dom0 is monitored based on the volume of packets passing through dom0. The monitored data are then weighted on the cost of each network operation. The packet processing cost is assigned according to the packet source and destination because, as shown in Table 5.1, the impact on dom0 CPU utilization of a packet depends on whether the packet comes from or goes to a domU or an external machine. If a router exceeds the allocated CPU, it is punished. Hence, to avoid attacks that generate unfair CPU punishments, it is important to define the responsible domain for each measurement operation. In transfers between domUs, the domU that sends the packet is responsible for all CPU costs to avoid an opponent domain from starting unsolicited traffic to exhaust CPU resources of a common domain. Besides, in transfers between domU and dom0, the cost of using CPU is always accounted for by the domU.

MUC controls memory usage by observing the size of the forwarding table of each virtual router. If the memory of dom0 reaches critical limits, all the virtual routers using tables or filters occupying more memory than the fixed reservation are punished, through the disposal of a percentage of routes. To avoid packet losses, a default route to the virtual router is added. Hence, packets that would match a discarded route are forwarded by the virtual router instead of discarded by dom0. As a result, reducing routing table sizes does not incur packet drops, but only in a reduced forwarding performance because the packet is forwarded by domU instead of by dom0.

5.3.1.1. *Punishment computing*

Opponent domains are punished by having their packets or routes dropped. MUC searches and converges for a dropping probability that balances the use of dom0 resources among virtual routers according to the fixed reservation and weight values. To avoid drops when there are idle resources on the physical machine, a virtual router is punished only if its usage surpasses its fixed reservation value and if the total resource utilization reaches a critical level, given by a percentage β of the total resources $R(t)$ in dom0. Without idle resources, all nodes that use more than their fixed reservation are punished to avoid other virtual routers to be harmed. Assuming that the total non-reserved resources are given by $D(t) = R(t) - \sum_{\forall i} \alpha_i R(t)$, then the dropping probability in $t + T$, given by $\Phi_i(t + T)$, is updated according to algorithm 5.1.

It is important to note that even if a domU consumes fewer resources than its fixed reservation value, the punishment is not immediately reset to avoid instabilities. Also, to prevent traffic generated by virtual routers to interrupt other dom0 services due to CPU overload, a residual punishment is constantly applied in the output interfaces of virtual machines. Such punishment should be small enough so as not to impact the low-volume transmissions but should prevent a domU consuming all dom0 resources.

5.3.1.2. *MUC prototype description and analysis*

The effectiveness of MUC is analyzed in the presence of opponent domains to verify the efficiency of the controller to share resources. A prototype was implemented in C and Python and the controller is able to monitor both bandwidth and CPU of dom0. Monitoring and punishment were implemented with `Iptables`. To dynamically estimate the capacity of the physical links, the Mii-tool is used. The CPU utilization is monitored with MUC by estimating the cost in dom0 of each network operation, including communication between domUs, from domU to dom0 and vice versa, from domU to an external machine and vice versa, and between external machines. The residual punishment in output virtual interfaces was estimated as 0.0009 based on the packet rate that severely impacted the dom0 response time.

Tests are run in a machine, hereinafter called router, equipped with an Intel Core2 Quad processor with 4 GB of RAM with five physical Gigabit Ethernet interfaces, using Xen 3.4-amd64 in router mode. All the four virtual machines instantiated run Debian operating system with Linux kernel 2.6 26-2, each one with a virtual CPU, 128 MB of memory and five network interfaces. The number of virtual CPUs in dom0 varies according to the test, and there are no memory constraints for this domain. The physical CPUs are shared among all virtual CPUs, and the hypervisor dynamically maps virtual CPUs onto real CPUs. The test uses two external machines to generate and receive packets, each with a network interface of 1 Gb/s. All traffic is generated with Iperf, and the results have a confidence interval of 95%.

---

input : $\Phi_i(t)$, W_i, α_i, $R(t)$, $U(t)$, $U_i(t)$, $D(t)$, β
output: $\Phi_i(t+T)$
1 **if** $(\alpha_i \cdot R(t) < U_i(t))$ *or* $(\Phi_i(t) > 0)$ **then**
2   **if** $(\alpha_i \cdot R(t) < U_i(t))$ **then**
3   % Calculate an idle resource usage indicator
4        $\Upsilon_i(t) = (U_i(t) - \alpha_i \cdot R(t))/D(t)$
5   **if** $(\beta \cdot R(t) \leq U(t))$ **then**
6   % Since there are no idle resources, some networks can be damaged. Thus, the punishment is increased.
7   **if** $(\Phi_i(t) > 0)$ **then**
8             $\Phi_i(t+T) = min(\Phi_i(t)+(1+\Upsilon_i(t))\cdot(1+\frac{1}{W_i})\cdot\frac{\Phi_i(t)}{(3-\frac{1}{W_i})}, 1)$
9   **else**
10             $\Phi_i(t + T) = \Phi_{initial}$ % Set initial punishment
11   **end**
12   **else**
13   % Reduce punishment, because there are idle resources
14           $\Phi_i(t + T) =$
            $max(\Phi_i(t) - (1 + (1 - \Upsilon_i(t))) \cdot (1 - \frac{1}{W_i}) \cdot \frac{\Phi_i(t)}{(3+\frac{1}{W_i})}, 0)$
15   **end**
16   **else**
17   % Reduce punishment, because the router used only its fixed resources.
18        $\Phi_i(t + T) = max(\Phi_i(t) - 3 \cdot (1 - \frac{1}{W_i}) \cdot \frac{\Phi_i(t)}{3}, 0)$
19   **end**
20 **else**
21     $\Phi_i(t + T) = 0$
22 **end**

Algorithm 5.1. *Heuristics for punishment computing.*

The first test evaluates the availability of the secure data plane update. This test is considered successful if the domU securely updates its data plane and no operation from other domains can prevent its completion. Thus, the impact of the MUC utilization during the data plane update is analyzed. The test consists of a maximum of three attempts from $domU_1$ to update the data plane, whereas $domU_2$ sends TCP traffic to $domU_1$. The scenario simulates an opponent virtual router, $domU_2$, trying to prevent a common router, $domU_1$, from operating normally. In MUC, $domU_1$, which is trying to update the data plane, has $\alpha_1 = 0.5$ and the opponent, $domU_2$, has $\alpha_2 = 0.3$. All domUs have weight $W = 500$.

Figures 5.3a) and b) show, respectively, the success probability of a data plane update and the volume of data transmitted between two virtual machines. An attack from $domU_2$ is effective when using the conventional plane separation, even if there is a great number of CPUs in dom0. MUC, however, increases the probability of a successful data plane update by up to 100%. Indeed, MUC limits the attack traffic from $domU_2$ avoiding the overload of dom0 resources. Also, MUC reserves the CPU resources required by $domU_1$ to send the update messages as well as performing cryptographic operations required by the secure plane separation. Figure 5.3b) shows MUC punishing traffic from $domU_2$ to $domU_1$ to ensure that the dom0 CPU resources are not exhausted. Therefore, the throughput achieved when using MUC is smaller than when using the secure data plane update with only one CPU in dom0. When the number of CPUs in dom0 increases, the CPU restriction is relaxed, and then the throughput using MUC increases. Indeed, MUC throughput is even greater than the throughput of the secure data plane separation. MUC ensures the fixed resources from $domU_1$, and then $domU_1$ can handle the data plane update as well as the ACK messages of the TCP connection started by $domU_2$. Losing ACK messages is worse for throughput than the limitation imposed by MUC to traffic because of CPU consumption. Thus, MUC ensures a secure data plane update with high availability and also ensures a high-performance connection between virtual machines as a result of the proposed architecture with the controller module.

The second test evaluates the transmission delay when using MUC with plane separation. The impact of MUC is evaluated compared with the usage of the plane separation paradigm without any control. This test measures the delay caused by MUC overhead according to dom0 workload. Because fairness is not evaluated in resource sharing, a virtual network is created with a fixed reservation of 100%. The test consists of two experiments that measure the round trip time (RTT) between two external machines using Ping. In the first experiment, there is no background traffic, whereas in the second experiment, background TCP traffic was generated between the two external machines. Results of both experiments are shown in Figure 5.4a). Without background traffic, data transmission presents a low RTT for both configurations. Nevertheless, when there is background traffic, the dom0 CPU is overloaded, increasing the response time of the system, and consequently, increasing the RTT. Results show that the CPU and the bandwidth control provided by MUC prevent dom0 CPU from becoming overloaded; thus, MUC presents an RTT up to eight times lower than the conventional plane separation configuration. It is important to note that even though MUC control incurs dropping packets, these drops do not cause a major impact on traffic, as shown in Figure 5.4b).

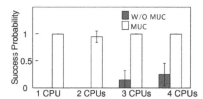 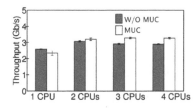

a) Probability of successful data plane update with traffic between $domU_1$ and $domU_2$

b) Throughput between $domU_1$ and $domU_2$ during data plane update

Figure 5.3. *Availability of the secure data plane update when using MUC*

The third experiment concerns sharing output links. To accomplish this, a domU and an external machine on different virtual networks initiate a communication with another external machine. Thus, both networks share the output link to the destination machine. Both networks have equal access to physical resources, with $\alpha = 0.5$ and $W = 500$ in both virtual routers. To assess MUC, the bandwidth control using traffic control (TC) is also tested, using the hierarchy token bucket (HTB). HTB creates two output queues, each one with a minimum bandwidth of 512 Mb/s and a maximum bandwidth of up to 1 Gb/s, to simulate the same resource usage policy than MUC. Figures 5.5a) and b) present the obtained results when domU sends User Datagram Protocol (UDP) traffic with a maximum rate of 1.5 Gb/s and packets of 1,500 bytes while the external machine sends TCP traffic. In the beginning, there is no control in the network.

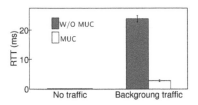
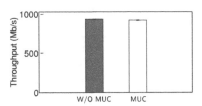

a) MUC impact over RTT between two external machines with one CPU on dom0

b) Background traffic throughput between two external machines with one CPU on dom0

Figure 5.4. *Impact of MUC over the RTT*

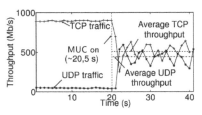
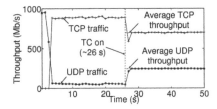

a) MUC with UDP traffic from domU and TCP traffic from external machine

b) TC with UDP traffic from domU and TCP traffic from external machine

Figure 5.5. *Resource sharing control using MUC*

Results show that external machine traffic has priority over domU traffic when there is no resource sharing control. This is an isolation failure when using plane separation because external traffic influences the maximum volume of the traffic generated by a domU. Thus, an external machine that belongs to an adversary network could inject traffic to damage the performance of a virtual router of another virtual network. In this experiment, the available resources are equally shared between the two virtual networks. Then, both networks should have an equal link slice, which means 512 Mb/s for each virtual network. Although MUC introduces a larger variation in traffic than TC, MUC average throughput has a lower error with respect to the ideal rate of 512 Mb/s for each virtual network than the TC average throughput. In fact, if there is no control of the external machine inflow, UDP traffic from the virtual machine is underprivileged and is unable to achieve high rates. Therefore, MUC control presents a maximum throughput error of -14.2% for UDP traffic and of -0.62% for TCP traffic with respect to the ideal rate of 512 Mb/s. TC, on the other hand, presents a maximum throughput error of -52.18% for UDP traffic and of $+35.68\%$ for TCP traffic with respect to the ideal rate of 512 Mb/s. This result shows that MUC presents a higher fairness in link resource sharing because it is adapted to Xen architecture particularities.

5.4. Summary

In this chapter, we explained a developed algorithm to control virtual networks' SLAs. The internal operation of this proposal is detailed as well as its interfaces to the piloting plane for controlling the virtual network environment according to the desired policies and primitives. Results show that it is possible to orchestrate the resources assigned to each virtual router and, furthermore, to each virtual network.

5.5. Bibliography

[BHA 08] BHATIA S., MOTIWALA M., MUHLBAUER W., *et al.*, Hosting virtual networks on commodity hardware, Technical report, Princeton University, Georgia Tech, and T-Labs/TU Berlim, January 2008.

[CAR 12] CARVALHO H.E.T., DUARTE O.C.M.B., "Elastic allocation and automatic migration scheme for virtual machines", in *Journal of Emerging Technologies in Web Intelligence (JETWI)*, Academy Publisher, no. 4 vol. 4, November 2012.

[EGI 07] EGI N., GREENHALGH A., HANDLEY M., *et al.*, "Evaluating Xen for router virtualization", *International Conference on Computer Communications and Networks (ICCCN)*, Honolulu, Hawaii, USA, pp. 1256–1261, August 2007.

[EGI 08] EGI N., GREENHALGH A., HANDLEY M., *et al.*, "Fairness issues in software virtual routers", *ACM Workshop on Programmable Routers for Extensible Services of Tomorrow (PRESTO)*, Seattle, WA, USA, pp. 33–38, August 2008.

[FER 11] FERNANDES N.C., DUARTE O.C.M.B., "XNetMon: a network monitor for securing virtual networks", *IEEE International Conference on Communications (ICC 2011 - Next Generation Networking and Internet Symposium (NGNI))*, Kyoto, Japan, pp. 1–5, June 2011.

[GEN 08] GENI PROJECT OFFICE, "Geni spiral 1 overview", September 2008. Available at http://groups.geni.net/geni/attachment/wiki/SpiralOne/GENIS1Ovrvw092908.pdf.

[GUI 13] GUIMARÃES P.H.V., FERRAZ L.H.G., TORRES J.V., *et al.*, "Experimenting content-centric networks in the future internet testbed environment", in *Workshop of Cloud Convergence: Challenges for Future Infrastructures and Services (WCC-02) - ICC'2013*, Budapest, Hungary, June 2013.

[HAN 09] HAN S.-M., HASSAN M.M., YOON C.-W., *et al.*, "Efficient service recommendation system for cloud computing market", *2nd International Conference on Interaction Sciences: Information Technology, Culture and Human (ICIS'09)*, Seoul, Korea, pp. 839–845, November 2009.

[HOU 10] HOUIDI I., LOUATI W., ZEGHLACHE D., *et al.*, "Adaptive virtual network provisioning", *Second ACM SIGCOMM Workshop on Virtualized Infrastructure Systems and Architectures (VISA 10)*, New Delhi, India, pp. 41–48, September 2010.

[JIN 09] JIN X., CHEN H., WANG X., *et al.*, "A simple cache partitioning approach in a virtualized environment", *IEEE International Symposium on Parallel and Distributed Processing with Applications*, Chengdu, China, pp. 519–524, August 2009.

[KEL 08] KELLER E., GREEN E., "Virtualizing the data plane through source code merging", *ACM Workshop on Programmable Routers for Extensible Services of Tomorrow (PRESTO)*, Seattle, WA, USA, pp. 9–14, August 2008.

[KOH 99] KOHLER E., MORRIS R., CHEN B., *et al.*, "The click modular router", *Operating Systems Review*, vol. 5, pp. 217–231, December 1999.

[KOU 01] KOUNAVIS M.E., CAMPBELL A.T., CHOU S., *et al.*, "The Genesis kernel: a programming system for spawning network architectures", *IEEE Journal on Selected Areas in Communications (JSAC)*, vol. 19, pp. 511–526, March 2001.

[MAT 12] MATTOS D.M.F., FERRAZ L.H.G., COSTA L.H.M.K., *et al.*, "Virtual network performance evaluation for future Internet architectures", in *Journal of Emerging Technologies in Web Intelligence (JETWI)*, Academy Publisher, no. 4 vol. 4, November 2012.

[MCI 06] MCILORY R., SVENTEK J., "Resource virtualisation of network routers", *IEEE Workshop on High Performance Switching and Routing (HPSR)*, Poznan, Poland, pp. 15–20, June 2006.

[MCK 08] MCKEOWN N., ANDERSON T., BALAKRISHNAN H., *et al.*, "OpenFlow: enabling innovation in campus networks", *ACM SIGCOMM Computer Communication Review*, vol. 38, pp. 69–74, April 2008.

[PIS 11] PISA P.S., COUTO R.S., CARVALHO H.E.T., *et al.*, "VNEXT: Virtual NEtwork management for Xen-based Testbeds", *2nd IFIP International Conference Network of the Future (NoF'2011)*, Paris, France, November 2011.

[SCH 09] SCHAFFRATH G., WERLE C., PAPADIMITRIOU P., *et al.*, "Network virtualization architecture: proposal and initial prototype", *ACM Workshop on Virtualized Infrastructure Systems and Architectures (VISA)*, Barcelona, Spain, pp. 63–72, August 2009.

[SHE 10] SHERWOOD R., CHAN M., COVINGTON A., *et al.*, "Carving research slices out of your production networks with OpenFlow", *ACM SIGCOMM Computer Communication Review*, vol. 40, pp. 129–130, January 2010.

[VIN 10] BAVIER A., FEAMSTER N., *et al.*, "In VINI veritas: realistic and controlled network experimentation", *ACM SIGCOMM Computer Communication Review*, New York, NY, USA, vol. 36, no. 4, pp. 3–14, August 2006.

[WAN 08] WANG Y., KELLER E., BISKEBORN B., *et al.*, "Virtual routers on the move: live router migration as a network-management primitive", *ACM SIGCOMM*, Seattle, WA, USA, pp. 231–242, August 2008.

[ZEC 03] ZEC M., "Implementing a clonable network stack in the FreeBSD kernel", *Usenix Annual Technical Conference (USENIX 2003)*, Santo Antonio, Texas, USA, pp. 137–150, June 2003.

Chapter 6

Piloting System

Autonomic networking [DOB 06, GAÏ 06, CHE 06] has been proposed to cope with the increasing complexity of communication networks, and it is a specific topic in autonomic computing [KEP 03], a term coined by IBM, which deals with systems' complexity by enabling their self-management. The aim is to release network administrators of the need to deal with tasks that require human intervention such as setting management policies and promoting automation of tasks such as system configuration and optimization, disaster recovery and security.

Self-management is also an essential element of the approach which advocates that the Internet of the future should adopt a pluralism of architectures [TUR 05]. Such pluralism indicates that network providers should be divided into service producers and infrastructure providers [FEA 07] and virtualization should be employed [AND 05] for that. In these networks, users request network services to the service providers, which instantiate virtual networks (VNs) over network substrates furnished by the infrastructure providers. Each VN can have its own protocols to assure the achievement of the objectives of the service running on it. Moreover, isolation of VNs is a requirement, i.e. the operation of a VN should not interfere with those of other VNs. To allocate resources to VNs in an optimal, robust and

Chapter written by Edmundo R.M. MADEIRA and Nelson LUIS S. DA FONSECA.

secure way is quite challenging due to the complexity needed to fulfill these requirements [ZHU 06, YU 08, HOU 08].

The Horizon project aims to define and validate a new network architecture based on pluralism and knowledge plane (KP) [CLA 03], with self-reconfiguration based on high-level instructions, automatic detection and correction of problems. To achieve such aims, it is necessary to have a piloting plane (PP) responsible for the decision-making process. This chapter reports the results obtained by the use of a self-management system for PP proposed in the Horizon project to decide on the use of control and management entities which consider the context of network and services.

The aim is to employ the PP in the design of the Horizon project, to handle VNs by using a multi-agent system. Piloting in the Horizon architecture is based on piloting agents (PAs), which can work in federations.

This chapter is organized as follows. Section 6.1 introduces some concepts in autonomic piloting systems (APSs). Section 6.2 discusses the PP functions and requirements. A preliminary PP design is presented in section 6.3, while section 6.4 shows the piloting agent architecture. Section 6.5 introduces the testbed used to validate the self-management system. The tools used for a first testbed and some preliminary experiments are also presented in this section. Section 6.6 presents the description and the multi-agent models of this preliminary self-management system. Section 6.7 reports the results of the final experiments. Section 6.8 presents a second self-management prototype multi-agent system, however, with a focus on self-healing of VNs. A set of experiments was performed in this context. Finally, section 6.9 draws some conclusions.

6.1. Autonomic Piloting Systems

APSs, as initially defined in the IBM manifesto [IBM 06], have been employed for the management systems of a single system, and they are supposed to perform different management tasks covering various nodes, links and services. The existence of different standards makes it impractical to devise a single autonomic control loop that autonomically adjusts different operational aspects such as fault, configuration, accounting, performance and security (FCAPS). This implies that there is a need to define one or more autonomic loops for each aspect so that the design of each control loop can be simplified. Moreover, the operation of the network is the result of the interaction of all control loops, which, however, might have conflicting goals. For example, an autonomic security component using a heavy encryption scheme to improve security can require too much processing and bandwidth, bounding the maximum network throughput to a level below the performance established by the service level agreement (SLA). In addition, if two sub-networks have conflicting configurations, either a renegotiation and reconfiguration process is required or translation services (gateways) must be installed in the border of the two sub-networks.

To solve such a problem, a new plane, the PP, is proposed to enable cooperation of the various autonomic control loops to ensure that decisions are not conflicting. Such cooperation (piloting) makes the overall optimization of the goals of autonomic component and control protocol aligned with the goals and SLAs defined for the whole network. Piloting also means that autonomic management domains employed by different operators are capable of automatically adjusting their configuration for accommodating federations of networks. Indeed, the need for a PP arises from the deployment of several autonomic control loops with different management goals, which would not be able to interoperate by themselves, to allow negotiation, federation and deployment functions. Thus, piloting deals with the meta-management of APSs, i.e. the deployment and reconfiguration of autonomic management control loops to allow their interoperation. This is achieved on the basis of a set of high-level goals, defined for the managed network domains that compose the network. Negotiation of new SLAs and policies, de-activation of conflicting management systems followed by the activation of others and migration of these systems within the piloted network may be needed to facilitate manageability of the network. The entire piloting process follows piloting policies, which dictate what the compromises are that each one of the managed domains should make for the sake of interoperability.

6.1.1. *Architecture*

The architecture of VN environments can be composed of four planes:

– The data plane for forwarding data.

– The control plane that contains all the control algorithms.

– The management plane (MP) responsible for all management features.

– The PP for feeding information, in real time, to the management and the control planes.

The data plane is responsible for transporting data from one sender to one or several recipients, which can virtualize elements in a network that employs virtualization. The management and control planes provide algorithms for failure detection, diagnostics, security management, routing, flow control, security control, mobility, etc. Indeed, different algorithms for the same task could be available at the nodes. The specific algorithm may depend on the context as well as on the need of the network and services.

A general view of such autonomic architecture is illustrated in Figure 6.1 that shows the introduced PP, which aggregates two specific sub-planes: the KP and the orchestration plane (OP). The KP should be able to quickly recover the knowledge which is useful for feeding the control and management algorithms, and the OP should coordinate (synchronize) network elements (NEs). These two planes cannot be dissociated in implementations, and therefore they need to be integrated. Moreover, the OP needs to have knowledge of the piloting intelligent process. A first implementation of this PP was proposed as a meta-control plane [GAÏ 96]. [GAÏ 06] and [BUL 08a] describe in detail some parts of this architecture and [BUL 08b] provides some examples. Another proposed change in relation to classical architecture is the fusion of the management and control planes in just one plane to get a three-plane autonomic architecture (Figure 6.2).

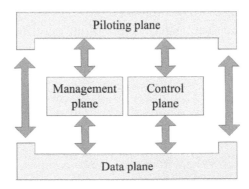

Figure 6.1. *A general view of autonomic architectures*

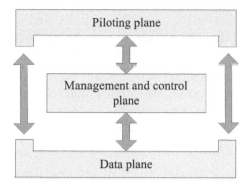

Figure 6.2. *The piloting-oriented architecture*

The PP drives the network using the control algorithms. For this purpose, the PP has to furnish information to the management and control algorithms. In summary, the PP has to orchestrate the management and control planes, which configures the data plane. For that, a distributed intelligent agents system allows the achievement of an adaptive control process because each agent holds different processes (behaviors and dynamic planner (DP)) allowing timely and most relevant decisions. Agents are implicitly cooperative in the sense that they use a situated view taking into account the state of the neighbors.

6.1.2. *Piloting plane of the horizon project*

The purpose of the PP is to govern and integrate the behaviors of the network control in response to changing context and in accordance with high-level goals and policies. It supervises and integrates all other planes' behaviors, ensuring integrity of management operations. The PP can be seen as a control framework into which any number of components can be plugged to achieve the required functionality. Moreover, the PP can be federated with other PPs. The PP also optimizes network monitoring and ensures that the required knowledge is available when required. The PP can use either locally available knowledge or global knowledge to manage long-term processes. The PP hosts several APSs, and it involves one or more PAs, and a dynamic knowledge base (KB) consisting of a set of data models and ontologies and appropriate mapping logic. Each APS represents a set of virtual entities, which manage a set of virtual devices or networks using a common set of policies and knowledge. The APSs access a KB that consists of a set of data models and ontologies. APSs can communicate and cooperate with each other, by the use of behaviors, which act as stubs for the APS communication. The PP acts as control workflow for all APSs ensuring bootstrapping, initialization, dynamic reconfiguration, adaptation and contextualization, and optimization. The PP supports the service lifecycle management, which is composed of creation, deployment, activation, modification and any operation related to the application services and/or management services. The APSs enables the following functions:

– *Federation*: allows domains (such as APS domain or piloted domain) to be combined into a larger domain (such as piloted domain or two-combined piloted domain) according to common high-level goals, while maintaining local autonomy. In APS Federation, each APS is responsible for its own set of real and virtual resources and services governed by it. In piloting federation, two piloted domains federate to make a larger piloted domain. In such cases, the federation should take into account the goals of the piloted domains to evaluate the feasibility of the federation.

– *Negotiation*: can happen between autonomous entities with or without human intervention. APSs and PAs are the main entities that can be engaged in negotiations. Each PA advertises a set of capabilities (i.e. services and/or resources) that it offers for the use of other components in the PP. APSs negotiate to support SLAs, defined by the operators of the managed piloting domains.

– *Distribution*: the APSs provide communication and control services that enable tasks to be split into others that can run concurrently on multiple PAs within the PP.

– *Governance*: each APS can operate in an individual, distributed or collaborative mode (i.e. in federation). The APS collects appropriate monitoring data in order to determine if physical or virtual resources and services need to be reconfigured. High-level goals, service requirements, context, capabilities and constraints should be considered as part of the decision-making process.

– *System Views*: APSs are responsible for managing the system views that are stored and diffused using the KP. APSs fetch the information required for their operation from the PAs as well as services and resources through interfaces.

6.1.3. *Related work*

D. Clark [CLA 03] suggests the construction of a new generation of networks able to "self-manage", given high-level objectives without any human intervention. Clark's proposal of a KP can be seen as a merge of the management, piloting and KPs of the Horizon project. Other autonomic architectures, such as Focale, proposed by Motorola [STR 06], extend the KP concept by introducing high level goals. Other architectures were proposed in the European FP7 program. They are:

1) The Autonomic Network Architecture (ANA) Project explores novel ways of organizing and using networks beyond legacy Internet technology [ANA 11]. The focus is mostly in protocols, not in management and piloting of networks.

2) The HAGGLE project deals with an innovative paradigm for autonomic opportunistic communication [HAG 11]. It develops a cross-layer network architecture exploiting intermittent connectivity by supporting opportunistic networking paradigm. In this project, the piloting is realized at the device level, while in the Horizon project it pilots the whole networks.

3) The BIONETS addresses the challenges of pervasive computing [BIO 11] by adapting dynamics of societies to deal with heterogeneity and to achieve scalability via an autonomic peer-to-peer communication paradigm.

4) The CASCADAS project has the objective of developing Component-ware for Autonomic Situation-aware Communications and Dynamically Adaptable Services [CAS 11]. It aims to propose an innovative architecture based on self-organized distributed components for autonomic and situation-aware communication.

5) The Ambient Networks (AN) [NIE 05] is an FP6 project that aims to develop a software-driven network control infrastructure for wireless and mobile networks that will run on top of all current network physical infrastructures. The objective is to allow devices to connect to each other and through each other to the outside world and to provide seamless service provisioning and roaming.

6) 4D is a new architectural model for the Internet, where tasks are divided into four planes: decision, dissemination, discovery and data [GRE 05]. In 4D, the data plane is a simple plane, which acts based on the configurations received by the decision plane. Decisions are made based on the information retrieved from the discovery plane, which constructs a view of the physical

resources. The decisions are then sent to the data plane using the dissemination plane. The paper presents neither simulation nor implementation to show the benefits of the architecture; however, the authors argue that the main advantage of the architecture is the centralization of decisions onto one single plane, removing the problems of multiple layers dealing with similar issues. 4D differs from Horizon by two main issues. First, 4D fuses the management, control and PPs into a single one. However, it does not describe a framework or design patterns to make the design of autonomic networks a tractable task. Second, 4D does not deal with the fact that we cannot rely on a single management entity because each domain is operated by a different organization. The PP, however, accounts for this fact, allowing the negotiation and federation of different management domains.

6.1.4. *Interaction of piloting, management and virtualization planes*

The Horizon autonomic management architectural model consists of a number of distributed management systems involving the four planes: virtualization plane (VP), MP, KP and PP. Together, these distributed systems form a software-driven network control infrastructure that will run on top of all current VNs and physical infrastructures to provide a way for devices to connect to each other, and through each other to the outside world as well as to provide seamless service provisioning. The PP will interact with the MP through behaviors, defined in section 6.3.2. Each APS will control one or more PAs, and each APS will deal with the piloting issues related to the interoperation of the PAs overseen by the PA. For supporting these tasks, the APSs will require information from the VP, using some interfaces to fetch the required information. Further, APSs will support service deployment, starting up or closing down network and user services. The APS defines constraints on the deployment of new services, such as the set of virtual routers or networks where the service will be installed as well as some of its execution parameters.

6.1.5. *Responsibilities of the piloting plane in the horizon architecture*

The role of the PP is to govern, dynamically adapt and optimize autonomic control loops in response to changing piloting-aware context and in accordance with applicable high-level goals and policies. It supervises and integrates all the other planes, behaviors, ensuring integrity of the management and control operations. Besides adapting the configuration of PAs, the PP may also bootstrap and close down PAs when needed. The PP can be thought as a control framework into which any number of components can be plugged to achieve the required functionality. The need for a PP arises from the deployment of several autonomic control loops with different administrators or management goals, which would not be able to interoperate without a set of translation, negotiation, federation and deployment functions. Thus, piloting deals with the meta-management of APSs, that is the deployment and reconfiguration of autonomic management control loops to allow their interoperation. This is achieved based on a set of high-level goals defined for each of the managed network domains that form the piloted network. The PP ensures the interoperation of management systems, even though those systems use a different set of high-level goals and management standards. Ontology translation and mapping techniques between different data models based on the common information model can be used to create the common language upon participating entities that can negotiate, federate, etc. This process of interoperation of management systems can be accomplished through negotiation of new SLAs and policies, deactivation of conflicting management systems followed by the activation of other management systems or migration of such systems within the piloted network. The entire piloting process is governed by piloting policies, which dictate what the compromises are that each of the managed domains are willing to make for the sake of interoperability.

6.2. Piloting plane functions and requirements

The PP collaborates with all the other planes and as such it demands the following functions and requirements:

1) Consideration of high-level goals and customer needs, excluding low-level technical and high-level business details.

2) Knowledge required for intra- and inter-domain piloting must be timely disseminated by the PP and related components of the architecture.

3) Reaction time of the APSs must be constrained within certain boundaries in order to achieve the defined SLAs.

4) Interfaces to interact with the PAs, that is to cope with environment changes and conflicts.

5) Support for PAs to define their dependencies on other components and services, as well as their expected operational conditions (e.g. based on policies describing their required services and virtual resources).

6) Cooperation of APSs from different domains requiring the use of open protocols and standardized information and data models.

7) Awareness of the state of virtual resources using the specific interface between VP and PP, and the information stored in the KP.

8) Solution of conflicts arising from orthogonal goals in different PAs. Thus, the PP must be capable of reaching compromises, allowing the system to achieve its purpose.

9) Workflow control for all PAs ensuring bootstrapping, initialization, contextualization and closing down of PAs. It should control the sequence and conditions in which one PA invokes other PAs in order to perform useful functions (i.e. a piloting is the pattern of interactions between APSs).

10) Enhancement and evolution: the PP should allow a relevant number of components to be plugged either into or out, to achieve the required functionality without interruption of normal operation.

6.3. Preliminary piloting plane design

The PPA concept in the Horizon approach is composed of several APSs. Each APS will be responsible for the interaction of several subordinated PAs. APSs will interact among themselves whenever necessary in order to guarantee end-to-end SLAs and service-level objectives (SLOs). The PP governs the execution of the APSs. It acts as a control workflow for all APSs ensuring bootstrapping, initialization, dynamic reconfiguration, adaptation, contextualization, optimization, organization and closing down of PAs. It also controls the sequence and conditions of one PA invoking other PAs to perform some useful functions (i.e. a piloting is the pattern of interactions between PAs). The DP is responsible for those tasks. Finally, each PA will interact with the KP to fetch information regarding the controlled behaviors (see section 6.4). This information will vary for each PA and will compose the situated view of the PA. The situated view defines the information that is needed for the operation of an autonomic component and from where it must be collected.

The PP is made up of one or more APSs, one per domain. Each controlled PA has its associated behavior. The behavior is a wrapper for the PAs, providing the interfaces and functionality needed for the interaction with the PA. There is one APS for each piloted domain, which communicates with the APSs of other piloted domains to reach agreements, allowing the operation of the network as a whole. A federation of the PP with other PPs is possible using communication between APSs. In this case, APSs from two or more administrative domains negotiate their federation. The aspects of federation, negotiation and governance will be treated in detail in section 6.3.2.

A policy-based management system abstracts the behavior and dynamic decisions of a system from its functionality. Therefore, the functionality of a system may stay the same, but its behavior, specifically with respect to changing contextual conditions, can vary. Policies are used in the PP to adapt its behavior with respect to changing contextual conditions, for example changing business objectives, network resource availability or domain membership. As the PP is concerned with the piloting of distributed management systems, its policies are directly associated with their requirements (PAs in the Horizon project). Essentially, the behavior of the components of the PP, namely the PAs, is dictated by the capabilities of the distributed PAs. For example, a single PA can be deployed using an APS. The behavior of this deployment may need to be dictated by the "owner" of the PA that will define the policies with which the PAs should be deployed, and the behavior of the APSs with respect to any coordination with the deployed PA. The APSs should enforce policies delegated to it by the PA for negotiation, federation or distribution. For example, a PA may inform its associated APSs that information regarding its resources should not be exposed to other PAs unless some strict authorization requirements are met. APSs can also control PAs that act directly on each VN element. One example for this organization is the handover process. A complete decision can be controlled by the PAs, while the definition of the values of the parameters for the handover decision (i.e. the maximum number of clients to be accepted into each VN and the authentication process being employed) can be determined by the high-level goals dictated by the APS.

Figure 6.3 illustrates the organization just described. The high-level APS controls the low-level PAs, setting the policies and SLAs for those components. The PAs act on the virtual resources using interfaces, deploying services and executing FCAPS functions of the virtual resources and nodes.

Finally, the low-level PAs control the real-time decisions required for the operation of control protocols. As defined earlier, the Horizon architecture has several levels of policies, continuously mapped by the piloting policies system. The PP uses piloting policies, which are high-level policies that control the deployment and federation of autonomic control loops. In a sense, piloting high-level policies are policies that act according to the inter-domain aspects of the autonomic management domains, controlling when and how two APSs can federate. They can also dictate how the process to resolve conflicts will be carried out by the APSs as well as the deployment and distribution of the PAs in the network. Meanwhile, the APS high-level policies will focus on the management of a single domain. The APSs receive the piloting high-level policies as well as the APS high-level policies from their corresponding administrative parties. Those policies are then processed, and relevant information from this processing is stored in the system view and can be accessed by the required components upon request. For example, the policies required by the federation of two APSs can be requested by a Federation Behavior, while the policies related to security can be requested by the self-security APS behavior. The design of the APS is outlined in Figure 6.4. It enables all components in the system managed by the PP to have plug-and-play behavior. APSs comprise three types of functional elements:

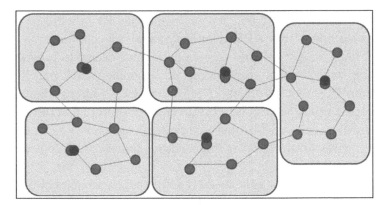

Figure 6.3. *Deployment of the APSs in the Horizon architecture*

– *Dynamic planner*: acts as a workflow engine for the execution of the behaviors in an APS. It is necessary to decide which behaviors have to be achieved.

– *Behaviors*: perform the specific/individual piloting actions required to be performed by an APS. They also represent specific management tasks in the network. The main behaviors are distribution, federation and negotiation. Behaviors are described in details in the following sections. Behaviors act as stubs for the PAs. They also perform internal functions specific to the APSs (core behaviors), for instance negotiation among behaviors.

– *Situated view*: is the "local window" of APSs into the KP. Views are composed of the intra-system view and inter-system view. The intra-system view provides an overall view of the system as seen by the components within a single APS. The inter-system view provides an overall, composite view of the APSs as a whole, that is it provides a view of the entire PP. The respective roles of these components are outlined in more detail further in the chapter.

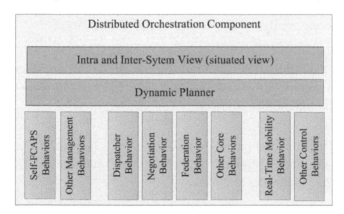

Figure 6.4. *Design of the autonomic piloting system*

6.3.1. *Dynamic planner*

The DP acts as a workflow engine for the execution of the behaviors. Such control includes the following tasks:

1) Definition of the sequence and conditions on which the behaviors must be bootstrapped for activating a PA (a deployment of the PA components that defines their dependencies on other behaviors; how and where to bootstrap them). It acts as a control workflow for all PAs ensuring bootstrapping, initialization, contextualization and closing down of PAs. It also controls the sequence and conditions under which one PA invokes other PAs in order to

carry out some useful functions (i.e. a piloting is the pattern of interactions between PAs).

2) Monitoring behaviors which incur checking that the operation of an individual behavior does not conflict with others.

3) Forwarding the advertised high-level policies for each piloted PA and core behavior, provided either by the operator or by the owner of the network.

4) Identification of conflicts in the initialization and reconfiguration of behaviors. The DP starts negotiation behaviors to reconfigure the deployment plan.

5) Triggering the initialization and closing down of behaviors with a plug-and-play, unplug-and-play approach;

6) Triggering the dynamic reconfiguration of behaviors;

7) Facilitating the interaction between behaviors. As an example of interaction, the negotiation behavior uses the DP to communicate with other behaviors in order to solve configuration conflicts. Then, it is up to the DP to trigger a reconfiguration behavior.

The DP will be configured by policies, which help the DP to select the behaviors (and hence services and/or PAs) that must be bootstrapped, the service or PAs parameters and SLAs.

The autonomic control loop of the DP (see Figure 6.5) is realized by the use of policies, which define rules based on events and conditions that must be met by the virtual resources. Whenever conditions are met, the DP executes the actions defined by the policy, which comprises the bootstrapping of one or more behaviors. Those policies can be changed on the fly, allowing the DP to reconfigure itself and to adapt to changes in the SLAs. Those policies are called high-level piloting policies in the Horizon architecture. More specifically, the DP will use the distribution policies of the piloting policies to identify where to instantiate or migrate PAs. Policies will also be used to decide which PA or core behavior should be bootstrapped and when. The APSs will also use policies to enforce SLAs and goals into the PAs.

The DP also acts as a mediator for the federation of the PAs. Whenever the DP enforces a new configuration on the PAs, they may refuse it due to self-governance. In this situation, the PAs will signal this rejection using the appropriate interfaces, and the DP will bootstrap a core behavior, responsible for conflict resolution, which will propose an alternative configuration.

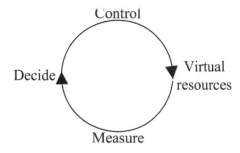

Figure 6.5. *General autonomic control loop of the dynamic planner*

6.3.2. Behaviors

Behaviors are sub-components of the APSs, which implement a certain piloting task. As an example, the APSs could implement behaviors for the negotiation of high-level policies, the distribution of tasks, the creation and destruction of services and virtual routers. Behaviors interact with each other when necessary, i.e. the federation behavior can interact with a Quality of Service (QoS) behavior if the required QoS cannot be met when two networks are joined. The lifecycle of a behavior is controlled by the DP, which identifies, starts and stops the behaviors necessary to accomplish a certain piloting task. We tentatively distinguish two types of behavior functional elements (FEs): the core behavior FEs, required for the proper operation of the PP, and the PA management behavior FEs, related to the control and management of the data plane. The core behavior supports the operation of the DP, implementing the tasks necessary for the proper cooperation of PAs and core behaviors. Some examples of core behavior FEs are listed below:

– *Distribution behavior*: sends and receives required data to different self-management piloting behavior FEs.

– *Negotiation behavior*: implements algorithms to mediate between the PAs in an APS, so that the PAs can agree on common goals. After the negotiation is complete, the PAs should take autonomous and local actions under their corresponding administrative domain, converging to these negotiated goals. Negotiation can also occur between different PAs.

– *Piloting knowledge update behavior*: manages the dissemination of knowledge regarding the PP. It controls information regarding the core behaviors required for the operation of the PP as well as the information required by the DP.

– *Network federation behavior*: controls the merge and separation of VNs controlled by different PAs. This behavior identifies the steps necessary to compose/decompose different federated domains, proposing actions to the DP.

– *Knowledge update behaviors*: supervises the operation of the KP. They define the "what, when and where". What information to collect, when to collect and from whom (where). These behaviors are specific to services; however, the whole set of behaviors supervises storage of information in the KP. Each PA requires pre-defined knowledge and runtime data. Mapping logic enables data, represented in a standardized information model, to be transformed into knowledge and combined with knowledge represented by ontologies.

– *Bootstrap and initialize behavior*: bootstraps and initializes other behaviors under the supervision of the DP.

– *Reconfiguration behavior*: dynamically reconfigures and adapts other behaviors under the supervision of the DP.

– *Optimizer behavior*: dynamically optimizes and organizes other behaviors under the supervision of the DP.

– *Closing down behavior*: dynamically closes down other behaviors under the supervision of the DP.

Examples of behavior FEs, which have direct interworking with other management functions providing real-time reaction, are listed below:

1) supervision of service lifecycle managers;

2) supervision of distribution/federation/negotiation of APSs;

3) supervision of interactions between APSs;

4) negotiation/distribution of the high-level goals to different APSs;

5) monitoring of the APSs;

6) supervision of network consistency/integrity checks of the sequence of changes to networks made by separate APSs.

In the following sections, we provide a detailed description of the main core management behaviors, which support federation, negotiation, governance and distribution of the management tasks. Next, we describe in more detail the PA behaviors.

6.3.2.1. *Federation core behavior*

Each APS is responsible for its set of virtual resources and services that it governs as a domain. Federation enables a set of domains to be combined into a larger domain and breaking down a domain into smaller domains. Federation behavior is important in finding a solution when conflicts between services are detected and an agreement cannot be reached as well as services not being able to find a resolution themselves. Federation is set up into three different attributions. The first case is when APS can federate the network by creating a domain by the merging of the two networks in conflict. The resulting domain inherits the characteristics of its components. In the second case, APS can federate by creating a domain that is the intersection of the federated networks. The new network requirements are those common to the two networks. In the third case, the APSs cannot federate by these two criteria. They create a bridge that enables the connection between the two networks.

There are two aspects of federation in the Horizon project. The first deals with the agreements among all of the federations participating, and the second facilitating the provisioning of services across federations. These two aspects are treated separately, but are dependent on each other because in case agreements cannot be made between participating members of the federation, there can be no consensus on service provisioning. One of the major challenges is that in the Internet, dynamic agreement adaptation between independent administrative domains is difficult when business and technical concerns need to be addressed individually. In such cases, the APSs should start a special negotiation behavior to help in the negotiation of a new agreement, which will be driven by the decisions of the APSs. In the Horizon project, a federation of APSs is separated according to either business or technical concerns, which is summarized in the following sections.

6.3.2.1.1. Federation of high-level objectives

A federation in Horizon is created when two or more APSs need to have a common objective. Typically, this objective is to provide a common set of services with guaranteed reliability across the boundaries of the APSs. Each

APS will contain its own business objectives as defined by the policy continuum. The APSs need to decide on a common set of business objectives that can be maintained across the federation. Once this set is established, the business aspect of the federation is addressed. Moreover, modifications of the business objectives can be proposed by any member. Any member can choose to leave the federation, leaving a federation is part of the self-governance function of the PAs and can incur a penalty if it breaches the terms of the federation. APSs can decide whether or not to participate in a federation, thus merely establishing some common understanding about how federation members should interact with each other. Actually, offering and consuming services in a federation is a technical issue and must be dealt with separately to allow new services to be introduced into a federation.

6.3.2.1.2. Technical concerns of a federation

Once an APS is participating in a federation, it can consume and provide resources and services in the federation. The APSs in the Horizon project are oriented toward the actual piloting of the usage of these services and resources and the joining and leaving operations of a federation. APSs must ensure that requested services are made available in a way that abides by the terms of the SLAs and policies specified by the federations. For that, the APSs must be able to evaluate if the configurations of services are adequate enough to ensure whether or not the agreed terms of the federation are upheld. The APS must also abide by its own business objectives and, if these objectives are no longer fulfilled, it may decide to leave the federation. On leaving a federation, the associated APS must signal this intention to the federation members.

Different types of services and resources require different technical solutions to ensure that they can be used effectively in a federation. It is up to the service creator to ensure that if the new service is to be used within a federation, it should be technically feasible to do so.

6.3.2.2. *Distribution core behavior*

The APSs provide communication and control services that enable tasks to be split into parts that run concurrently on multiple APSs within a PP, or even across multiple PPs.

6.3.2.2.1. Business concerns of distribution

The business concerns relate to whether or not each APS requires the distribution of information, tasks and code. Information distribution is important, as some APSs may generate information that is unique to them and will need to be explicitly controlled. Access control policies can help with that. Task distribution can be seen as the distribution of work that involves distributed processing across a number of processing elements. For example, the computation of the effective bandwidth of a specific video can be distributed across a number of machines because this can be a time-consuming process. The business concern, in this case, refers to the fact that a specific APS may request its associated PAs to distribute a processing task. The distribution of code is slightly different from the distribution of information because tasks typically expect no return from the distributing PA. The distribution of code can be an enabler for an APS to upgrade a specific service offering that relies on the federation of a number of PAs.

6.3.2.2.2. Technical concerns of distribution

The distribution of information can be carried out using the concepts in the KP related to the context information service. However, the APSs need to be involved as they can enforce strict access control on the information distribution to abide by the high-level policies of the APS. The high-level policies can indicate whether information can leave or enter the system. The distribution of tasks can technically be carried out by the platform being used to distribute information. However, the task information can be in a specific format, in which strict instructions may need to be in place to instruct the target PAs of the requirements of the task. It is up to each PA to decide whether to accept or reject the task being distributed. The PA needs to inform the APSs about its intentions. The APS then takes care of the distribution and informs the PA whether or not the intentions were met.

The distribution of code can also be carried out by the platform being used to distribute information. Executable code in this case is a special form of information that can be distributed. The code can be used to upgrade or deploy a new service to other PAs. Therefore, the PA instructs the APSs that the code needs to be distributed and it is up to the APSs to handle the distribution of the code. This can be carried out using the service enabler plane (SEP).

6.3.2.3. *Negotiation core behavior*

In some approaches for the Internet of the future, it is mandatory for network and service providers to offer and publish their services so that more complex services can then be provided. An important component that allows this requirement to be met in the Horizon project is the negotiation component of the PP. This component acts as a virtual service broker that mediates between different PAs, taking care of service requests and providing support so that the underlying service providers can negotiate responsibilities, tasks and high-level goals. This support should be carried out taking into account the nature of the underlying service providers, the services they provide, their interests, their service qualities and other key aspects. All in all, this functionality should be provided, leveraging the service requester entities from complex decision-making processes.

In the Horizon project, each APS advertises a set of capabilities (i.e. services and/or resources) that it offers for use in the PP. The negotiation component enables a specific functionality of selected capabilities to be agreed among APSs. Examples include using a particular capability from a range of capabilities (e.g. a particular encryption strength when multiple strengths are offered), being granted exclusive use of a particular service when multiple APSs are competing for the sole use of that service and agreeing on a particular protocol, resource and/or service to use. The negotiation functionality piloted between APSs and PAs has inherent business and technical concerns. The following section elaborates on these two critical aspects.

6.3.2.3.1. Business and technical concerns for negotiation

When PAs negotiate high-level goals under an APS, or when different PAs negotiate high-level goals with other PAs, the negotiation finishes when the participants align their internal business objectives with a common objective. That is to say when the participants converge to the negotiated high-level goals for which virtual and physical resources must be allocated, managed and controlled autonomously in their respective domains (PA and/or APS). In the negotiation of business objectives, the responsibilities, benefits and penalties are also considered during the negotiation.

It is worth mentioning that an APS and/or a PA can negotiate business objectives with several PAs and APSs, sequentially or in parallel. The negotiation of business objectives is, in turn, influenced by technical concerns in the sense that APSs compromise resources to fulfill the negotiated high-level goals. The governance capability of an APS (and that of a PA) defines the reason, and when, to negotiate. The negotiation capability ensures that APSs and PAs can always negotiate with other entities in a federation.

PAs and APSs can have active negotiated agreements with several parties. There is a potential need to renegotiate the high-level goals due to statistical changes in the resources committed to these goals when they cannot be fulfilled or due to some internal decisions that lead to such corrective actions.

The APSs can also trigger renegotiation when common business objectives are not fulfilled. Renegotiation of high-level goals is a functionality that should be supported by the PP. It is a part of the governance capability of each PA and/or APSs to decide on which of its negotiated business objectives to renegotiate. The renegotiation can be driven by utility functions, cost/benefit optimizations, etc. Again, during the renegotiation of high-level goals, the responsibilities, benefits and penalties are also considered.

The PP provides the means for the mediation between PAs and APSs so that they can negotiate high-level goals as a result of a complex service request. As the PAs and APSs may belong to different administration domains, they may talk different languages and may express their high-level goals in different terms. Ontology translation and mapping techniques can be used to create the common language upon participating entities. The PP must provide mechanisms for the negotiation to occur regardless of any of these technical aspects.

6.3.2.4. *Piloting core behavior*

The governance functionality of an APS deals with the self-interested actions that it takes to (re) negotiate high-level goals with other parties and to take actions that can involve the commitment of virtual and physical resources that can help provision of services across federations; each PA can operate in an individual, distributed or collaborative mode. In each case, the PA collects appropriate monitoring data in order to determine if the virtual and non-virtual resources and services that it governs need to be (re) configured. Business objectives, service requirements, context, capabilities and constraints are all considered as part of the self-interested decision-making process. APSs can also be federated with other APSs (inter-system situated view) because they can act in self-interest, namely they can have self-governance properties.

6.3.2.4.1. Business and technical concerns for governance

In a federated environment, the APSs provide support in such a way that their underlying PAs are aware of the needs of other PAs within a federation.

APS can decide which PAs it will collaborate with. As introduced earlier, each APS can operate in an individual, distributed or collaborative mode:

– The APSs act as individual entities, when they are not part of any federation, working isolatedly and autonomously, and governing its own virtual and physical resources and services which are driven by its own business objectives. Neither common goals nor responsibilities are shared with other APSs. APSs should handle and transmit the messages sent by the corresponding PA to other PAs and/or APSs in the federation in which they are located.

– As a result of the distribution function of the PP, APSs can work with other APSs in a federation where complex services are provisioned by coordinating the activities, resources and services of each distributed APS after a negotiation phase. Each APS should provide a number of contract interfaces to the PAs, which use them to promote and mediate the negotiation of a complex service, its activation and maintenance. These interfaces could be interpreted as contract interfaces that isolate the internal structure and capabilities of the PAs.

– An APS works in a collaborative mode when it acts as a local or a global collaborator. A local collaborator is responsible for coordinating the functionality of other APSs in a given PP, when the APSs delegate part of their control to an APS. An APS works as a global collaborator when it coordinates the functionality of APSs across different PPs, namely among APSs. This enables the emulation of client-server, n-tier, clustered and peer-to-peer architectures.

Regardless of the operation mode, the PP should provide the means for an APS to govern its virtual and non-virtual resources through a well-defined set of interfaces. The aim of the APSs is to allow the PAs to always decide on the action to take based on self-interested policies, driven by business objectives and capabilities. Its decisions should also take into account the policies of other APSs participating in the federation. However, the APS can reject abiding by certain policies of the federation when those are not aligned with its own business objectives. In this case, it is up to the APSs to renegotiate or change the policies of the federation.

6.3.2.5. *APS behaviors*

The APSs use the APS behaviors as an interface to communicate with PAs. Such encapsulation allows the APSs to see the PAs in a uniform way, having the same interfaces as the core behaviors. Communication between PAs and APSs follows a defined information model. Further, APS behaviors are wrappers to the CPA, similar to stubs and skeletons in Remote Procedure Calls (RPC) or CORBA, once they provide a translation from the specific design issues of an APS to the operational philosophy of the APSs. One of the uses of such a wrapper is to hide different implementations of the APS from the APSs, for example ensuring a single communication point, even though the APS can be a distributed component spread around several virtual nodes.

The wrapper also defines a set of commands that the APSs must support to allow the APSs to orchestrate their federation and distribution. The interface provides mechanisms for the PAs to disseminate and renegotiate policies to the APSs, allowing the PA to be self-governing. The PAs can also support near real-time control of virtual resources and protocols. While the high-level self-FCAPS APSs are mainly concerned with long-term management of resources, low-level PAs are deployed to react as fast as possible to changes in the PA-aware context. Thus, low-level PAs use simple algorithms that act based on a predetermined overall goal. Those goals are dictated by policies defined by the APS. Those PAs will act over a single virtual resource or a small set of virtual resources due to time constraints on their reaction. In a sense, low-level PAs can be seen as the first control loop of an autonomic system, acting based on predetermined goals and with no sort of embedded learning. Examples of possible technologies that could be used are state machines, proportional-integral-derivative (PID) controllers, fuzzy decisions, etc. Low-level PAs are associated with the maintenance of a QoS defined by the network SLA, controlling the parameters of the virtual nodes and their running algorithms, for example mobility management algorithms.

6.3.3. *Intra- and inter-system views*

APSs use the KP to store and disseminate the information required for their operation. The information can be decoupled into two parts, or views, according to their relevance to a given APS.

The intra-system view concerns information required to orchestrate the services within the piloting domain, whereas the inter-system view deals with the piloting of several piloting domains. The intra-system view contains information that enables APSs to become aware of the specific situation. The inter-system view provides similar information for collaborating APSs.

The intra-system view thus deals with the services running within the boundaries of an APS with information relevant to the operation of the APS. This view is also called a situated view. Indeed, the situated view can be definitely different depending on the algorithm to be fed. For example, not only a one-hop or a two-hop or a complete situated view can be defined, but more complex situated views can also be defined, as shown in Figure 6.6.

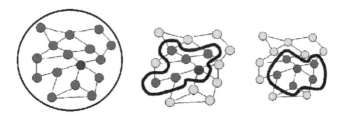

Figure 6.6. *Different solutions for the situated view*
(entire, intermediate and one-hop)

There is an important trade-off on the definition of the situated view: large situated views allow decisions to be made using a description of the overall state of the system. However, the cost of disseminating and processing, such knowledge among all the concerned nodes, is high. Thus, an optimal situated view should be defined based on the problem being solved and on clear performance metrics.

APSs maintain the situated views, defining which information must be stored in these views, for which virtual nodes or resources, and how frequently it must be updated. The interface of the PP with the KP is described in the following section.

6.3.4. *Interfaces of the APS*

This section describes the interactions of APSs with other elements of the Horizon architecture, as shown in Figure 6.7. We first describe the interface of the APSs with the KP. The interfaces with virtual resources, the APSs and the service enabler's plane are introduced here.

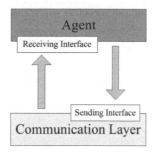

Figure 6.7. *Interfaces of the APS*

The KB is a repository in which all information and knowledge required for management and control tasks are kept. As each element should be able to work autonomously, the PP should also be able to provide the information required autonomously. The PP provides the APSs with the required parameter values used for monitoring and from where in the network information should be gathered. These parameter values are decided by the dynamic planner, based on the functionality of PP and on the state of the network, and these values raise a specific behavior dedicated to this task. Two types of knowledge in the KB should be considered: those related to management tasks and those related to piloting tasks.

One of the objectives of the PP is to identify information that is required for autonomic decisions, and next to activate the required KP functions that will ensure the timely collection and delivery of this information to the APSs or to a specific behavior. The following functions are required for the interaction between KP and OP:

– *Lookup(Info)*: requests the lookup of a certain piece of information (or knowledge) from the KP. This is used for fetching policies, SLAs as well as context and relevant configuration parameters of the piloted APSs.

– *Store(Info)*: stores information on the KP. This function is used for knowledge produced in the OP, which is then stored on the KP.

– *Subscribe(Event, Component)*: is defined by a condition on the stored information of the KP, this function allows the APSs to be notified of relevant events happening in the network. Events can define conditions and sets of nodes or APSs on which they can happen. For example, APSs can be interested in watching for the occurrence of certain faults to trigger the reconfiguration of an APS responsible for fault management. The subscription also identifies which APS and which of its components will process the information.

– *Watch(Info) and unWatch(Info)*: are used by the APSs to define which information must, or not, be periodically collected and disseminated by the KP, respectively. For example, the averaged bandwidth utilization of a network interface associated with a certain PA can be monitored to verify if this component should be migrated to another node in the network. The information fetch is defined by the type of information, its situated view (from which virtual nodes must be collected), the periodicity of such an update and the component requiring this information. Once the APSs issue a *Watch* request, the KP is responsible for the maintenance of the information, delivering it directly to the concerned component. One example of use of *Watch* is when an APS defines the number of active flows that should communicate to the autonomic performance optimization PA. When this information is no longer required, the APSs use the *unWatch* function to release the KP resources.

– *Push(Information, PAs)*: is used for synchronous messages passing among the PAs. The *Subscribe*, *Watch* and *Push* interfaces of the APSs are used by them to create their inter-system view, used to feed the dynamic planner and the core behaviors with their correct parameter values. Those functions are also used for inter-APS communication because an APS can watch the state of other APSs using the same functions. The access to information from other APSs should be controlled by access policies because APSs can be controlled by different organizations, and thus APSs might wish to hide some aspects of the management of their networks.

– *Interface SEP/APS, Virtual Resources/APS and PA/APS*: As described earlier, the ultimate task of the PP is to use PAs on the network. The deployment of PAs should be done considering the management requirements of services. The PP performs PAs deployment, using the SEP. However, PP uses the SEP to use PAs, it does not deal with the deployment of services. The PP needs to interact with virtual resources to provide the PP with physical management and control information on the state of the network for the deployment of PAs behaviors. The last interface used by the APSs is the interface of the APSs with the PAs. This interface is indeed implemented by the PA wrapper behaviors.

6.4. The piloting agents

In this section, we describe the distributed piloting agent plane. This plane is composed of distributed intelligent agents each associated with a NE, or a set of NEs, as illustrated in Figure 6.8.

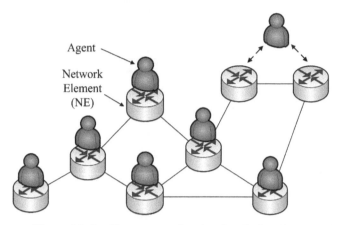

Figure 6.8. *Intelligent agents forming the piloting plane*

By distributing agents across the network, the PP allows us to deal locally with problems. Indeed, local problems are often simpler and easier to deal with than those resulting from remote issues. Furthermore, a local problem can be addressed earlier locally than in a centralized approach; for example, an agent can immediately change the configuration of its NE to react to a local load problem. Beyond purely local problems, agents can cooperate to deal with problems in this neighborhood, for example a connectivity problem can be detected by several agents. Agents can then cooperate to characterize the problem more precisely and eventually provide a solution or a report to the higher levels. Agents can use powerful distributed AI techniques to perform their tasks. The KP which is a part of the PP consists of the situated view component of the agents. Similarly, the OP is composed of a dynamic planner and different behaviors. The architecture of the piloting system is illustrated in Figure 6.9.

The first part of the PP is composed of a KP which is a distributed information base embedded in agents. It contains local information of the agent as well as global information of the network. Each agent maintains its own view of the network on the basis of information obtained directly from local observation of its NEs, as well as indirectly from the rest of the network by exchanging information with its neighbors. This agent-centric view of the network, focused on the close network environment, is called the situated view and is illustrated in Figure 6.10.

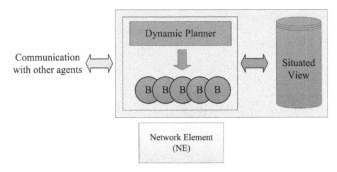

Figure 6.9. *Outline of agents architecture*

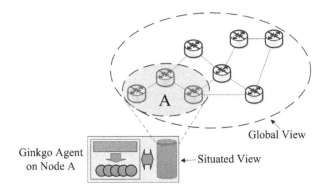

Figure 6.10. *Each agent has its own situated view of the network*

The rationale for the situated view is that events occurring in the neighborhood of an agent are generally of greater importance for the agent than events occurring in a remote part of the network. The fact that local events are known earlier and are more accurately documented in the situated view makes it easier for the agent to react rapidly and appropriately. Agents regularly check for important changes appearing in their situated view and in the network environment. It can decide to automatically adapt certain parameters of their managed NEs or ask neighboring agents to do so for their respective NEs. The use of the situated view drives implicit cooperation between agents which "influence" each other by the knowledge that they are sharing. Implicit cooperation is the primary mode of cooperation among agents in the PP. This mode of cooperation is simple, particularly robust and well suited for dynamically changing environments because it does not require the establishment of an explicit dialogue and a strict synchronization between agents.

An autonomic system is composed of different autonomic elements, which cooperate to achieve the overall objective of the autonomic system. Therefore, an OP is required to decide, in real time, what to do when a threshold is reached. The OP is responsible for the process of government and integration of the behaviors of the system in response to changing context, and in accordance with applicable goals and policies [TSE 09]. The autonomic architecture performs this process via behaviors and dynamic planner. What an agent is capable of doing is defined as a set of behaviors ("B" in Figure 6.11). Each of these behaviors can be considered as a specialized function with some expert capabilities to deal with specific aspects of the work to be performed by the agent.

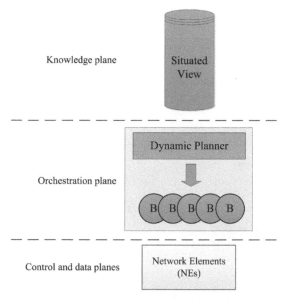

Figure 6.11. *Outline of the piloting system*

Typical categories of behaviors are listed in the following:

– Updating the situated view in cooperation with other agents.

– Reasoning individually or collectively to evaluate the situation to decide on the application of appropriate action, for example a behavior can simply be in charge of computing bandwidth availability on an NE, it can also regularly perform a complex diagnostic scenario or it can be dedicated to automatic recognition of specific network conditions

– Acting on the NE parameters, for example a behavior can tune QoS parameters in a DiffServ context.

Behaviors can access the situated view that operates within each agent as a whiteboard shared among the agent's behaviors. The activation, dynamic parametrization and scheduling of behaviors within an agent is performed by the dynamic planner. The dynamic planner decides which behaviors have to be active, when they have to be active, and with which parameter values. The dynamic planner detects changes in the situated view and occurrence of external/internal events. From there, it orchestrates the reaction of the agents to change in the network environment.

6.5. Testbed

It is important to evaluate the proposed APS before deploying the system in a real network. Figure 6.12 illustrates a testbed built with this objective [SEN 11].

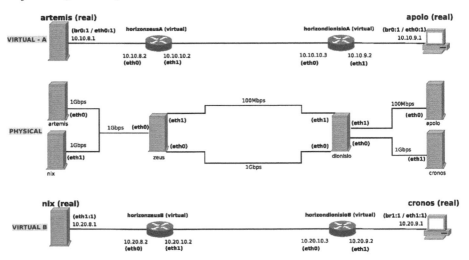

Figure 6.12. *testbed*

The testbed contains two VNs; VN A (VIRTUAL-A, Figure 6.12a)) and VN B (VIRTUAL B, Figure 6.12b)). Both VNs contain two virtual routers each. The virtual routers are located at the real hosts *zeus* and *dionisio*. To the VN A to be instantiated, it is needed to instantiate the virtual routers *horizonzeusA*, on the real host *zeus*, and *horizondionisioA*, on the real host *dionisio*. Similar instantiations are needed to the VN B.

VN A is created to interconnect hosts *artemis* and *apolo* through a two-hop virtual path. Similarly, VN B is created to interconnect hosts *nix* and *cronos*. The virtual links between the virtual routers can be mapped either on an 100 Mbps link or an 1 Gbps link between the physical routers *zeus* and *dionisio*. Initially, the two virtual paths share the 100 Mbps link.

Preliminary experiments were performed to confirm the possibility of changing the mapping of the virtual paths during the operation of both VNs. The idea was to overload the 100 Mbps link and, after that, to migrate just one virtual path to the 1 Gbps link. This experiment simulates a scenario in which PAs located at the routers detect the high utilization of a link and make the decision to migrate one of the virtual links.

The following sections present more details about the testbed. Section 6.5.1 summarizes some tools used to build the testbed, and section 6.5.2 presents the results of preliminary experiments. It is important to note that these experiments were realized manually without mediation of PAs from the proposed APS. The implementation of the multi-agent APS is commented on in section 6.6, and the results obtained with the PAs are presented in section 6.7.

6.5.1. *Tools*

In this section, we describe the major tools used to build the testbed used in the experiments. The tools are used for the deployment and manipulation of VNs, and the Ginkgo platform was used for the development of the multi-agent APS.

6.5.1.1. *Qemu*

qemu [BEL 05] is a processor emulator that can also be used to create virtual machines (VMs). Compared to other virtualization platforms, qemu has the advantage of being easy to install because, to the host operation system, it is an application like any other. The disadvantage is that qemu is susceptible to the same process scheduling and memory management algorithms of the host operating system.

Qemu is a piece of free software under the General Public License (GPL) and open source. There are some extra components that can be used with qemu to improve the performance of the guest operating system when the real processor contains the same instructions of the emulated processor. The idea is to allow the guest operating system to access the real processor directly.

6.5.1.2. *KVM*

The kernel-based virtual machine (KVM) [KIV 07] is a full virtualization hypervisor based on the machine emulator qemu. It runs on x86 architectures with new virtualization technologies like Intel VT and AMD-V. KVM consists of a kernel module of Linux, which, when loaded, provides an interface at /dev/kvm to the initialization and control of the guest VMs.

KVM is a free software under the GPL and open source, that allows us to use external tools to control it, like libvirt.

Some of the KVM features that are interesting for the Horizon project include:

– good performance in full virtualization;

– live-migration of VMs;

– support SMP hosts and guests;

– memory ballooning;

– VM networking by bridging, routing or private networks.

KVM surpasses qemu in terms of performance because the main objective of qemu is to be a processor emulator, not a virtualization platform. However, qemu has the advantage of running on any architecture, and is easy to install because there is no need to modify the host operation system with special kernels or modules.

Preliminary experiments of the multi-agent APS presented in this section were realized in a testbed powered by qemu VMs. The final experiments of the multi-agent APS, presented in section 6.7, were performed on a current testbed with KVM VMs.

6.5.1.3. *Libvirt*

Libvirt [COU 10] is an API to access the virtualization capabilities of Linux with support to a variety of hypervisors, including qemu, KVM and Xon, and some virtualization products from other operating systems. It allows local and remote management of VMs. With libvirt it is possible that a PA uses the same code to monitor and control virtual resources independent of the virtualization platform.

The libvirt is implemented in C (supporting C++) and includes direct support to Python. It also supports a number of language bindings, which have been implemented for Ruby, Java, Perl and OCaml. In the experiments, the C version and the Java binding of the API were used.

The importance of libvirt for this work is to make virtualization technology-independent, facilitating future design changes. With libvirt it would be possible to change the virtualization platform from qemu to KVM without modifying the PAs.

6.5.1.4. *Ginkgo distributed network piloting system*

Ginkgo distributed network piloting system [GIN 08] is an agent platform based on autonomic networks. It has building blocks for the development of a piloting system for computer networks. The framework allows the creation of lightweight and portable agents, which facilitates its implementation in heterogeneous environments: routers, switches, hosts, wired and wireless networks. The agents play the role of the autonomic manager of autonomic computing. With distributed managers close to its managed elements, monitoring can be done locally.

The platform allows the creation of clusters of agents. Neighboring agents exchange information and get a situated view of the network. This information is stored in the KB that uses an information model to facilitate communication between agents. The policy file is another repository, which contains rules for the application. With knowledge of the environment and application, the multi-agent system can provide a self-knowledge property to network.

The sensing, cognition and acting of agents are performed by the behaviors. They feed the KB, perceive and predict threatening events and perform changes on the managed elements. Agents also have a dynamic planner that can change parameters of the behaviors and control the lifecycle of the agent, based on information in the KB and on rules in the policy file. This makes possible the development of self-configuration, self-healing, self-optimizing and self-protection functions, promoting self-management.

The Ginkgo platform provides the key features necessary to implement the PAs, which are presented in section 6.4.

6.5.2. Experiments in the testbed

Preliminary experiments were performed to evaluate the possibility to change the mapping of the virtual links during the operation of the VNs. This section describes the results obtained in these experiments.

Virtual routers are VMs based on qemu version 0.12.5. For the creation and manipulation of them, the `libvirt` version 0.8.3 was used. Both the physical and VMs use the operating system Debian GNU/Linux with kernel version 2.6.32. The virtual links are created on the data link layer, using the Ethernet protocol (802.3), via virtual interfaces and bridges. The bridges are controlled by the utility `brctl` of the Bridge-util package, version 1.4-5.

Initially, the virtual links of VNs shared the 100 Mbps link of the substrate network. User Datagram Protocol (UDP) traffic was generated using the `iperf` in the two VNs until the 100 Mbps link was saturated. At this point, scripts were executed manually on the routers of the VN.

Figure 6.13 plots the round-trip time (RTT) between the two hosts *cronos* and *dionisio* connected by the VN B (VIRTUAL B in Figure 6.12) using the saturated 100 Mbps link. The scripts changing the mapping of the virtual link from 100 Mbps to 1 Gbps were executed at time 40 s. It is possible to observe that this change had, as consequence, the reduction of the RTT value between the two hosts. At time 60 s, the scripts were executed to return the mapping of virtual link to the 100 Mbps link. As consequence, the RTT increased to its initial range of values.

An experiment similar to the experiment summarized in this section is described in section 6.7. The difference is that the latter was performed with PAs taking actions, instead of the manual execution of scripts.

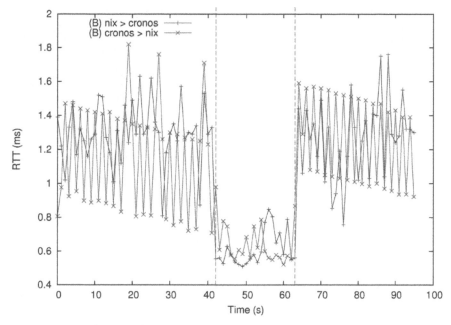

Figure 6.13. *RTT between hosts of the virtual network B*

6.6. The multi-agent APS

We implemented a multi-agent APS within the testbed to perform the management of VNs [SOA 12]. The PAs run on the nodes of the network, as depicted in Figure 6.14. PAs in the APS should monitor and control physical and virtual resources. The APS was developed with support from Ginkgo platform, version 2.0.13. The PAs were compiled and executed with the version 1.6.0-21 of Java. For the monitoring of VNs by the PAs, the libvirt-java library version 0.4.6 was used.

In the KB, each network resource follows an information model. Each resource is controlled by a single PA and, therefore, each representation of this resource is unique in the KB. In the information model, the links are directed, that is each wire is represented by two links in opposite directions. The PAs are located in routers, and they send data over a direct link, monitor and control the link. Figure 6.15 shows the class diagram of the information model, based on [FAJ 10].

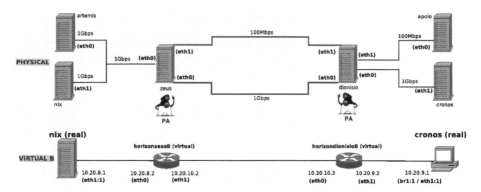

Figure 6.14. *PAs in physical routers managing virtual routers*

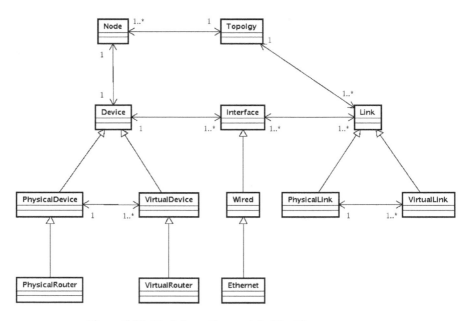

Figure 6.15. *The information model of the PAs*

The neighborhood consists of all PAs of the routers within the domain. All communication in the Ginkgo platform is based on the diffusion of information stored in the KB. The KB stores the data (topological information, status of the node, etc.) used by PAs in a homogeneous and structured way and provides diffusion mechanisms of this knowledge among PAs. The view that a PA has of its environment (situated view concept) can be directly implemented with the KB. Inside a PA, the KB can be seen as a common blackboard which can be accessed by the behaviors. Those behaviors implement autonomic processing and work as organic components permanently adapting themselves to environment changes. The overall function of these components is orchestrated by the dynamic planner, which follows a policy provided by the application programmer. The dynamic planner manipulates structured knowledge stored in the KB.

In this implementation, the PAs have four behaviors: *monitor*, *analyze*, *plan* and *execute*, which are executed periodically in sequence, forming the autonomic cycle of the manager. Their actions are listed as follows:

– *Monitor*: collects data from network interfaces through `libvirt` and feeds the KB.

– *Analyze*: evaluates the use of physical network links. If the rate of utilization exceeds the threshold defined in the policy, the physical link information is updated in the KB to inform the other behaviors.

– *Plan*: plans the action when the analysis behavior indicates a threat, that is it chooses another physical link, if any, to receive a flow of a virtual link that is mapped onto the overloaded link.

– *Execute*: executes the actions and informs the changes for other behaviors and PAs through the KB.

The policy file has a set of rules that are evaluated in every PA cycle (Figure 6.16). The rules have a name and a conditional expression. The rule RINIT (line 2) uses the *impulse init* condition, which is true once at the beginning of the PA execution. In this rule, the behaviors are configured and started with the *start* primitive. Following the policy file, it is also possible to set and get values in the KB. The first action in rule RINIT is to set the value in the *speed* property of the *control* individual object as 1.0 (line 3). This value represents the current period of behaviors' cycle in seconds. This period is defined with the *changerate* primitive. The sequence of behavior calls could be controlled by their priority, defined with the *changeprio* primitive. The behaviors could be parameterized with the *setcontrol* primitive. The dynamic planner applies the policy and sets the *threshold* attribute to 0.5 (line 7). This value represents a percentage use of the physical links. The rule RFASTER (line 17) changes the period of behaviors cycle to 0.2 s when the *Analyze* behavior turns on the *speedup* flag and the current *speed* is 1.0. The rule

RNORMAL (line 23) returns this cycle period to 1.0 s, when the *Analyze* behavior turns off the *speedup* flag and the current *speed* is 0.2.

```
[frame=lines,fontsize=\small,numbers=left]
(policy (subgoal main (rules
  (rule RINIT if (impulse init) (
    (set control.speed 1.0)
    (changerate MonitorBehavior 1.0)
    (changeprio MonitorBehavior 3)
    (start MonitorBehavior)
    (setcontrol AnalyseBehavior.threshold 0.5)
    (changerate AnalyseBehavior 1.0)
    (changeprio AnalyseBehavior 2)
    (start AnalyseBehavior)
    (changerate PlanBehavior 1.0)
    (changeprio PlanBehavior 1)
    (start PlanBehavior)
    (changerate ExecuteBehavior 1.0)
    (changeprio ExecuteBehavior 4)
    (start ExecuteBehavior)))
  (rule RFASTER if (and control.speedup (= control.speed 1.0)) (
    (set control.speed 0.2)
    (changerate MonitorBehavior 0.2)
    (changerate AnalyseBehavior 0.2)
    (changerate PlanBehavior 0.2)
    (changerate ExecuteBehavior 0.2)))
  (rule RNORMAL if (and (not control.speedup) (= control.speed 0.2)) (
    (set control.speed 1.0)
    (changerate MonitorBehavior 1.0)
    (changerate AnalyseBehavior 1.0)
    (changerate PlanBehavior 1.0)

    (changerate ExecuteBehavior 1.0))))))
```

Figure 6.16. *The rules of the policy file*

6.7. Results

We evaluated the time and packet loss of the automated adaptation on the VNs according to the changing environment, performed by the multi-agent APS. The actions are similar to those presented in section 6.5.2. However, in this experiment, the virtual routers are VMs based on KVM version 0.12.5. For the creation and manipulation of them, it was used the libvirt in version 0.8.3. Both the physical and VMs contain the operating system Debian

GNU/Linux with kernel version 2.6.32. The virtual links are created in the data link layer via virtual interfaces and rules in virtual switches inside physical nodes. We use Open vSwitch [PFA 09], version 1.3.0, as virtual switch and `ovs-ofctl` utility to control the flows.

The configuration parameters of the PAs are shown in the policy file (Figure 6.16). The behaviors of the autonomic cycle run at an interval of 1.0 s in normal state, when the utilization of all physical links in the network is below the *threshold* set in the *Analyze* behavior (configured at 50%). When an overload occurs, the cycle frequency increases to a rate of 5 runs per second, that is an interval of 0.2 s.

Figure 6.17 shows the utilization of the physical links throughout the experiment. A UDP flow, generated by the `iperf` application, passing through the VN *A* at a constant rate of 70 Mbps, starts at 5.0 s. The Fast Ethernet physical link reaches 70% of utilization, exceeding the *threshold* defined on the *Analyze* behavior, and the correction begins to be performed by the *Plan* behavior. The PA takes about 1.0 s to recognize the situation and adapt the network.

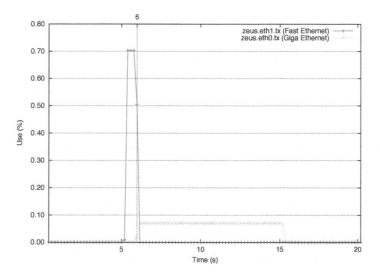

Figure 6.17. *Utilization of the physical links*

Figure 6.18 shows the traffic on VN *A*. As shown, the adaptation did not impact VN throughput. It means that the migration of virtual link to a new path does not incur packet loss. In this scenario, the virtual link between virtual nodes *horizonzeusA* and *horizondionisioA* was mapped onto a single physical link, between physical nodes *zeus* and *dionisio* (see Figure 6.14). If there were other substrate nodes on the path, the multi-agent APS would have to make the adaptation backward on the path to ensure no packet loss.

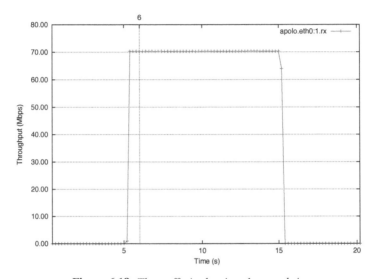

Figure 6.18. *The traffic in the virtual network A*

6.8. Multi-agent system for self-management of virtual networks

In this section, we show how autonomic computing techniques can be applied to the management of VNs. It presents a distributed architecture to enable self-management of VNs over a substrate network that makes part of the PP. The autonomic managers of the NEs maintain a closed control loop of monitoring, analysis, planning and execution, which feeds a KB for the next iterations. The focus is on the self-healing of VNs, but the distributed architecture for self-management of VNs is sufficiently generic to be used by other functionalities in autonomic computing such as self-configuration, self-optimization and self-protection. Related work is also introduced in what follows to contextualize the main contribution of this work.

A multi-agent system for maintaining SLAs in events of resource failure and severe performance degradation is presented in [HOU 10]. The agents form and manage groups based on similarity of physical nodes. The similarity function is also used to choose where the virtual nodes will be recovered in case of failure. Our architecture is also based on the multi-agent systems, and fast recovery mechanisms of VNs are implemented by restoring the virtual routers from backup memory to reduce the convergence time of the routing protocol.

In [MAR 10], a distributed management architecture is presented with the goal of self-organizing VNs to maintain effective utilization of physical resources. The algorithm used for self-organization is based on autonomic control loop. It monitors the target link, and it tries to minimize the traffic load in the network through the migration of virtual nodes. We present similar architecture and implement a prototype with focus on the self-healing of VNs to evaluate our proposal.

6.8.1. Implementation of the prototype

Figure 6.19 illustrates the testbed built to test the system in a controlled environment. The network core consists of four machines: *zeus*, *atlas*, *dionisio* and *cronos*. The machines *zeus* and *dionisio* are also connected to the hosts *apolo*, *hermes*, *nix* and *artemis* by Giga switches. On the substrate network, two VNs with three virtual routers were created. The images of virtual routers are in a repository accessible by all machines of the substrate network via Network File System (NFS). Both the physical and the VMs have the operating system Debian GNU/Linux with kernel version 2.6.32.

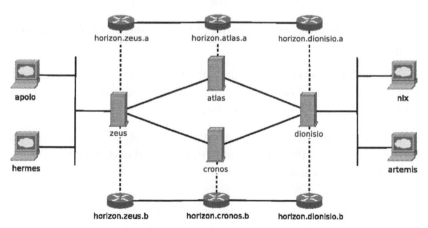

Figure 6.19. *Testbed built to validate the multi-agent system for self-management of virtual networks*

The management of VMs for both the creation of the testbed and the agents' operation uses the Libvirt library [COU 10]. It provides an API for monitoring and control of various virtualization platforms, including Xen. The use of Libvirt in the system for self-management of VNs allows it to be independent of the virtualization technology adopted.

The agents run on physical machines at the network core: *zeus*, *atlas*, *dionisio* and *cronos*. We use the Ginkgo platform [GIN 08] for the implementation of the agents. It allows the creation of lightweight and portable agents, which facilitate their deployment in heterogeneous environments. Ginkgo is a framework that has the basic building blocks for our architecture.

The multi-agent system must perform disaster recovery of VNs. Failures can occur by problems in virtual routers, in the physical nodes or even in the physical links. In the first case, the agent responsible for the virtual router must diagnose and notify others about the failure. In other cases, the agent can also stop communicating, and therefore the neighbors must diagnose the failure.

An autonomic control loop for self-healing of VNs was implemented. The loop is controlled by the dynamic planner and is formed by four behaviors: Monitor, Analyze, Plan and Execute. Agents regularly perform these behaviors in this sequence. They are described as follows:

– *Monitor*: collects data from virtual and physical nodes and feeds the KB.

– *Analyze*: performs fault diagnosis in virtual routers or physical nodes in the neighborhood.

– *Plan*: calculates the cost of the substrate node based on the use of its resources. The physical node with more virtual routers running on it has the highest cost. It also disseminates this information to other agents.

– *Execute*: verifies whether all agents in the neighborhood have already sent their information. If so, the agent in the physical node with the lowest cost recovers the failed virtual routers.

We create a special agent running on a host machine. This agent is part of the neighborhood and receives the other agents' information. It executes a behavior to create and update a graphical interface. This agent contributes to prototype an architecture with hybrid agents with different capabilities and levels of cognition. In this case, it performs visualization of the topology and the mappings of VNs over the substrate, and charts with the information collected by the *Monitor* behavior of the other agents. The graphical interface is shown in Figure 6.20.

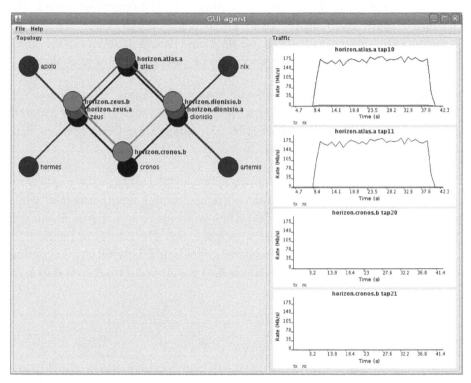

Figure 6.20. *Graphical interface of the multi-agent system*

6.8.2. *Experimental results*

The experiments aim to validate the architecture for self-management of VNs and test different approaches to recover virtual routers in cases where there are failures of physical elements. The recovery time T_r from a VN can be set according to:

$$T_r = T_d + T_p + T_i + T_c \hspace{3cm} [6.1]$$

where T_d is the time of failure diagnosed; T_p is the time spent on the planning action, which involves exchanging of information between agents; T_i is the time of instantiation of the VM and T_c is the convergence time of the routing protocol.

We studied two ways to recover the virtual router. The first creates a VM from the image file in the repository, booting the guest operating system. The second resumes the machine from the backup memory file, also in the repository. In this case, there is no time involved with booting the guest operating system and the convergence time of routing protocol is small.

The executions were performed with static and dynamic routing in the VN. Static routing is manually configured in the virtual routers. For the dynamic routing, the Quagga routing suite [ISH 13], running the Open Shortest Path First (OSPF) algorithm, was used.

The first experiment was performed in the testbed without the multi-agent system. Migrations of a virtual router with static routing were performed, while a UDP flow at constant rate of 500 Kbps from *apolo* to *nix* passed through the VN. The traffic was generated with Iperf tool. The curves in Figure 6.21 show the flow rate; a zero value indicates losses in the network. All migrations change the mapping of the virtual router *horizon.cronos.b* from *cronos* to *atlas* and began near 8 s.

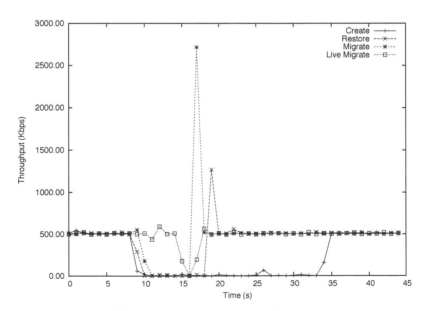

Figure 6.21. *Experiments without the multi-agent system*

In the first run, curve "Create" in Figure 6.21, the virtual router was destroyed at the physical router *cronos* and recreated at the physical router *atlas*. In this case, the recovery took 27 s mainly due to the time of booting the operating system. In the second run, curve "Restore" in Figure 6.21, the state of the virtual router was saved in the network file repository, the virtual router was stopped at the physical router *cronos*, saved in the image repository and restored at the physical router *atlas*. Even spending more time with the memory saving, the process took only 45% of the time taken in the first experiment. The third runs use migration services offered by KVM. In the basic migration, executed in the third run, curve "Migrate" in Figure 6.21, the VM is stopped, the memory is copied over the network from origin to destination, and then the machine is restored at the destination. This situation is similar to the second run, except that the copy is made directly from the physical router *cronos* to the physical router *atlas* and not through an intermediate machine, and this leads to a time that is 37% of the time taken in the first run. In these two situations, when the machines were restored there was a peak in the network. This happened because at the instant that the VM memory was saved, there were packets in the buffer that had been added to those that were arriving at the instant that it was reactivated. The fourth run, curve "Live Migrate" in Figure 6.21, was carried out with live migration, in which the memory is copied from source to destination while the VM was running, and only when there were no more modified pages was control transferred. This way, migrating a virtual router took less time, because the transmission was interrupted for the shortest period, which was less than 15% compared with that of the first run.

The curves in Figure 6.22 show the results of the experiment with the distributed management system and UDP traffic generated with Iperf tool. Figure 6.22a) shows the executions with static routing, and Figure 6.22b) with dynamic routing. In this figure, the approaches to create the virtual router from the image file and to restore the VM from the backup memory file are compared. In all runs, a UDP flow at constant rate of 20 Mbps from *apolo* to *nix* was sent through the virtual network B, and the curves plot the flow rates over time.

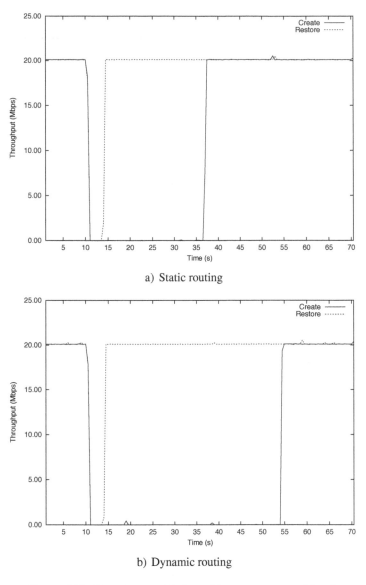

a) Static routing

b) Dynamic routing

Figure 6.22. *Experiments with the multi-agent system and Iperf*

After 10 s, the physical router *cronos* is disconnected from the network. Agents in *zeus*, *atlas* and *dionisio* diagnose the failure of *cronos* because they stop receiving the KB propagation of its agent. In the experiment, the agents perform the *Monitor* behavior and the KB dissemination every 0.5 s as well as the information about one physical node is outdated longer than 1 s. Therefore, the time of diagnosis of failure varies between 1 and 1.5 s. Thereafter, the agent enters in the *Plan* behavior, when it calculates the physical node cost that will be sent to other agents in the next propagation. The *Execute* behavior waits for all agents to send their physical node costs. The costs of *zeus* and *dionisio* are higher because they run two virtual routers. The agent in *atlas* recovers the *horizon.cronos.b*; *atlas* creates the VM from image or restores from the backup memory.

Another experiment with Secure Copy Protocol (SCP) application over VN was performed, and the results are shown in Figure 6.23. The SCP generates Transmission Control Protocol (TCP) traffic. A 1 GB file was transferred from *hermes* to *artemis* using the VN B. For each scenario, 10 runs were performed and confidence intervals were derived with a 95% confidence level. All executions were carried out under similar network conditions.

In this experiment, executions were performed without the agents to serve as a basis for comparison. First, the time to send the file through Path A (*zeus* → *cronos* → *dionisio*) and through Path B (*zeus* → *atlas* → *dionisio*) was measured. As shown in Figure 6.23, the flow rates of Path A and Path B are different. Thus, live migration of *horizon.cronos.b* from *cronos* to *atlas* was performed, 10 s after starting the SCP transmission. This is used as a basis for comparison, floor result, because live migration has the lowest downtime. Next, we execute the experiment with the agent system to perform the recovery of the VN. In these scenarios, a failure occurs in physical node *cronos*, also after 10 s. Again, the multi-agent system chooses *atlas* to recover *horizon.cronos.b*. Four scenarios with failure were performed, combining the routing schemes: static and dynamic, and the virtual router recovery approach, either creation of the VM from image or restoration of the VM from backup memory.

There is a considerable difference in the recovery time of the VN depending on the way the virtual router is recovered. The difference is greater when using dynamic routing, due to the convergence time of the routing algorithm. The recovery by creating the VM took 43.6 s in this case, and 26.4 s with static routing in the Iperf experiments with the multi-agent system. The time to complete the SCP transmission was 52% and 122% higher on average when compared to the floor value in the static and dynamic scenarios, respectively. The creation of VMs in dynamic routing scenario had a larger variance given the routing protocol convergence time.

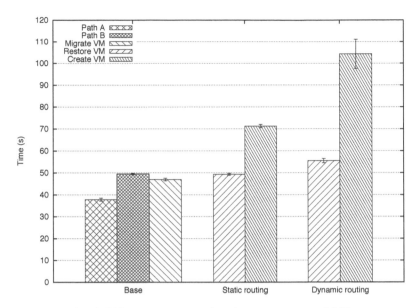

Figure 6.23. *Experiments with multi-agent system and SCP*

In the approach which the VM state is restored from backup memory, the type of routing had a lower impact. In both the static and the dynamic scenarios, the recovery time was about 3.5 s in Iperf experiments with the multi-agent system. The impact on the SCP transmission time was about 5% and 18% when compared to the floor value in static and dynamic scenarios, respectively.

Results show the approach of restoring the VM from memory backup, especially when dealing with virtual routers to decrease the convergence time of the routing protocol. Although this approach is faster, saving transmissions and storage, the backup memory can cause significant impact on network performance. This backup could occur in moments of network idleness to cause less impact on its operation. Distributed storage and memory ballooning can also be used to reduce this overhead.

6.9. Summary

The Horizon project proposes a new system or plane, the PP, which enables the cooperation of the various autonomic control loops in the network. The PP governs and integrates the behaviors of the network control doors in response to changing context and in accordance with high-level goals and policies. This chapter presented the initial design of the PP in the Horizon project, defining its functions, requirements and the concepts behind the operation of the components. The PP hosts several APSs, and it involves one or more PAs and a dynamic KB consisting of a set of data models and ontologies and appropriate mapping logic.

A PA is a functional entity of the Horizon project architecture that deals with inter-domain management tasks, such as federation, negotiation, governance and distribution of management domains. The APSs have two main building blocks. The first is the dynamic planner, which serves as an autonomic policy-based dispatcher. The dynamic planner creates and destroys behaviors, and implements the interfaces for the PAs and the core inter-management functions. The second building block represents a behavior, which has two functions. The first, carried out by the core behaviors, implements piloting tasks. The second, implemented by the PA behaviors, is to act as a proxy for the communication of the APSs with the PAs that it orchestrates.

In addition, we showed some results from an APS prototype, based on multi-agents provided by the Ginkgo platform to realize the PAs.

The idea of an APS to manage VNs can also be applied to a cloud environment. Figure 6.24 illustrates some ideas that can be explored in this direction. This figure shows a VN consisting of a group of five VMs running tasks in a Cloud environment (Instance of CloudRequest), the group of VMs being hosted on physical machines connected to the network substrate (VM-Server-Park). The network management is done by PAs that continuously monitor the resources. Each resource, whether physical or virtual, has at least one PA responsible for the monitoring and execution of actions. PAs can be grouped into domains (neighborhoods). The figure shows two neighborhoods of PAs acting jointly. A neighborhood consists of the PAs that monitor the VN (Instance of CloudRequest) and request actions to PAs in the other neighborhood, which monitor the network substrate (VM-Server-Park). An example of action can be a request to increase the bandwidth of the VN. Figure 6.24 also illustrates other types of PAs that assist in the task management of the cloud, such as power management, and management based on SLA.

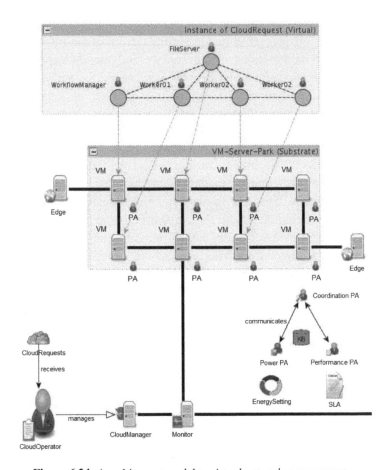

Figure 6.24. *A multi-agent model to virtual network management*

Finally, we presented a self-management system prototype, in which the concepts of autonomic networks were applied in a virtualized environment by the use of a multi-agent system. We evaluate the autonomic self-management environment in a scenario that demanded self-managing failures (self-healing). This infrastructure was used to build the prototype.

6.10. Bibliography

[ANA 11] ANA P., "Autonomic network architecture", 2011, available at http://www.ana-project.org/.

[AND 05] ANDERSON T., PETERSON L., SHENKER S., *et al.*, "Overcoming the internet impasse through virtualization", *Computer*, vol. 38, pp. 34–41, 2005.

[BEL 05] BELLARD F., "Qemu, a fast and portable dynamic translator", *Proceedings of the Annual Conference on USENIX Annual Technical Conference*, ATEC 05, USENIX Association, Berkeley, CA, pp. 41–46, 10–15 April 2005.

[BIO 11] BIONETS P., "Bio-inspired service evolution for the pervasive age", 2011, available at http://www.bionets.eu/.

[BUL 08a] BULLOT T., KHATOUN R., HUGUES L., *et al.*, Merghem-Boulahia L., "A situatedness-based knowledge plane for autonomic networking", *International Journal of Network Management*, vol. 18, pp. 171–193, 2008.

[BUL 08b] BULLOT G.P.T., GAÏTI D., ZIMMERMANN H., "A piloting plane for controlling wireless devices", *Telecommunication Systems*, vol. 39, pp. 195–203, 2008.

[CAS 11] CASCADAS P., "Component-ware for autonomic situation-aware communications, and dynamically adaptable services", 2011, available at http://acetoolkit. sourceforge.net/cascadas/.

[CHE 06] CHENG Y., FARHA R., KIM M.S., *et al.*, "A generic architecture for autonomic service and network management", *Computer Communications*, vol. 29, pp. 3691–3709, 2006.

[CLA 03] CLARK D.D., PARTRIDGE C., RAMMING J.C., *et al.*, "A knowledge plane for the internet", *Proceedings of the 2003 Conference on Applications, Technologies, Architectures, and Protocols for Computer Communications*, SIGCOMM 03, Karlsruhe, Germany, pp. 3–10, 25–29 August 2003.

[COU 10] COULSON D., BERRANGE D., VEILLARD D., *et al.*, "Libvirt 0.7.5: application development guide", 2010, available at http://libvirt.org/guide/pdf/Application_Development_Guide.pdf (accessed in January 2013).

[DOB 06] DOBSON S., DENAZIS S., FERNÁNDEZ A., *et al.*, "A survey of autonomic communications", *ACM Transactions on Autonomous and Adaptive Systems*, vol. 1, pp. 223–259, 2006.

[FAJ 10] FAJJARI I., AYARI M., PUJOLLE G., "VN-SLA: a virtual network specification schema for virtual network provisioning", *International Conference on Networking*, Paris, France, pp. 337–342, 11–16 April 2010.

[FEA 07] FEAMSTER N., GAO L., REXFORD J., "How to lease the Internet in your spare time", *SIGCOMM Computer Communication Review*, vol. 37, pp. 61–64, 2007.

[GAÏ 96] GAÏTI D., PUJOLLE G., "Performance management issues in ATM networks: traffic and congestion control", *IEEE/ACM Transactions on Networking*, vol. 4, Issue 2, pp. 249–257, 1996.

[GAÏ 06] GAÏTI D., PUJOLLE G., SALAUN M., *et al.*, "Autonomous network equipments", *Autonomic Communication*, LNCS – Lecture Notes on Computer Science, vol. 3854, pp. 177–185, 2006.

[GIN 08] GINKGO NETWORKS, "Ginkgo distributed network piloting system", *White Paper*, 2008, available at http://www.ginkgo-networks.com/IMG/pdf/WP_Ginkgo_DNPS_v1_1.pdf (accessed in January 2013).

[GRE 05] GREENBERG A., HJALMTYSSON G., MALTZ D.A., *et al.*, Refactoring network control and management: a case for the 4D architecture, Technical report, 2005.

[HAG 11] HAGGLE P., "A content-centric network architecture for opportunistic communication", 2011, available at http://code.google.com/p/haggle/.

[HOU 08] HOUIDI I., LOUATI W., ZEGHLACHE D., "A distributed and autonomic virtual network mapping framework", *4th International Conference on Autonomic and Autonomous Systems*, ICAS 08, pp. 241–247, March 2008.

[HOU 10] HOUIDI I., LOUATI W., ZEGHLACHE D., *et al.*, "Adaptive virtual network provisioning", *Proceedings of the 2nd ACM SIGCOMM Workshop on Virtualized Infrastructure Systems and Architectures*, VISA 10, ACM, New York, pp. 41–48, 2010.

[IBM 06] IBM, "An architectural blueprint for autonomic computing", *Autonomic Computing White Paper*, 4th ed., June 2006.

[ISH 13] ISHIGURO K., "Quagga: a routing software package for TCP/IP networks", 2006, available at http://www.quagga.net/docs/quagga.pdf (accessed in January 2013).

[KEP 03] KEPHART J.O., CHESS D.M., "The vision of autonomic computing", *Computer*, vol. 36, pp. 41–50, 2003.

[KIV 07] KIVITY A., KAMAY Y., LAOR D., *et al.*, "KVM: the linux virtual machine monitor", *Proceedings of the Linux Symposium*, vol. 1, pp. 225–230, 2007.

[MAR 10] MARQUEZAN C.C., GRANVILLE L.Z., NUNZI G., *et al.*, "Distributed autonomic resource management for network virtualization", *IEEE/IFIP Network Operations and Management Symposium*, NOMS 10, Osaka, Japan, pp. 463–470, April 2010.

[NIE 05] NIEBERT N., "Ambient networks: a framework for mobile network cooperation", *Proceedings of the 1st ACM Workshop on Dynamic Interconnection of Networks*, DIN 05, ACM, New York, pp. 2–6, 2005.

[PFA 09] PFAFF B., PETTIT J., AMIDON K., *et al.*, "Extending networking into the virtualization layer", *Proceedings of the 8th ACM Workshop on Hot Topics in Networks (HotNets-VIII)*, New York, October 2009.

[SEN 11] SENNA C.R., BATISTA D.M., SOARES M.A. JR, *et al.*, "Experiments with a self-management system for virtual networks", *Proceedings of the 2nd Workshop de Pesquisa Experimental da Internet do Futuro (WPEIF 2011)- XXIX Simposio Brasileiro de Redes de Computadores e Sistemas Distribuidos*, SBRC 11, Brazilian Computer Society, Brazil, 2011.

[SOA 12] SOARES M.A. JR, MADEIRA E.R.M., "A multi-agent architecture for autonomic management of virtual networks", *Proceedings of the 4th IEEE/IFIP International Workshop on Management of the Future Internet*, ManFi 12, IEEE Computer Society, Maui, USA, 16 April 2012.

[STR 06] STRASSNER J., AGOULMINE N., LEHTIHET E., "Focale: a novel autonomic networking architecture", *Proceedings of the Latin American Autonomic Computing Symposium*, LAACS 06, Campo Grande, Brazil, 18–19 July 2006.

[TSE 09] TSELENTIS G., DOMINGUE J., GALIS A., *et al.*, *Towards the Future Internet, A European Research Perspective*, IOS Press BV, 2009.

[TUR 05] TURNER J., TAYLOR D., "Diversifying the internet", *In Proceedings of the IEEE Global Telecommunications Conference, GLOBECOM*, IEEE, St. Louis, USA, pp. 755–760, December 2005.

[YU 08] YU M., YI Y., REXFORD J., *et al.*, "Rethinking virtual network embedding: substrate support for path splitting and migration", *SIGCOMM Computer Communication Review*, vol. 38, pp. 17–29, 2008.

[ZHU 06] ZHU Y., AMMAR M., "Algorithms for assigning substrate network resources to virtual network components", *INFOCOM 2006, Proceedings of the 25th IEEE International Conference on Computer Communications*, Barcelona, Spain, April 2006.

Chapter 7

Management and Control: The Situated View

The network virtualization technique allows the execution of multiple virtual networks over the unique physical substrate. This functionality is achieved through the use of a control and management entity, which multiplexes hardware access and provides logical slices of resources to the virtualized systems. The main primitives of a virtualized networking system are the creation and destruction of virtual networks, the migration of virtual nodes and the mapping of virtual networks on the physical substrate.

In the standard version of Xen and OpenFlow, all of the above-mentioned primitives are manually run, which implies scalability and management problems. Hence, we proposed and developed a piloting plane that is able to autonomously execute these primitives. We also developed a knowledge plane, which works together with the piloting plane, to detect when virtual networks are no longer working as desired and, consequently, require modifications on their properties by executing the primitives. The knowledge plane monitors virtual routers, obtains their usage profile, and, by using prediction mechanisms, proactively detects the necessity for updates in the virtual network configuration. The knowledge plane stores information concerning each virtual network element, allowing management decisions and proactive network maintenance. Because of the scalability issues, the knowledge plane is distributed in different nodes. Therefore, each node keeps

Chapter written by Otto Carlos M.B. DUARTE.

a partial view of the knowledge plane, limited to their neighborhood and the surroundings. This partial view is called the situated view of a node. The main challenges of the knowledge plane are the time scheduling to take decisions and the network element information updates. We define and select the information that must be stored on the knowledge plane, and also on which nodes this information will be stored. Finally, we define the update frequency for each type of information based on the performance and Quality-of-Service (QoS) metrics. We only proccess situated view information instead of whole network information due to scalability.

We developed a set of distributed mechanisms to detect and correct virtual network anomalies. The mechanisms identify and store network state variations and also predict variable evolution. The proposed mechanisms fit well for Xen and OpenFlow platforms, but we only describe here the mechanism for Xen platform.

The first proposed approach is based on the development of a dynamic allocation system that analyzes the resource utilization profiles of virtual routers and provides a fair share of resources based on QoS metrics and service level agreement (SLA). The second proposed approach focuses on monitoring and prediction techniques that monitor the environment, provide proper and updated information to the knowledge plane and also detect router misbehavior. These two approaches together constitute a framework for detecting the need for updates and also when SLAs are violated, which may also trigger update alarms.

This chapter is organized as follows. Section 7.1 describes the mechanism to extract and store virtual network profiles. Also, we describe a QoS controller implemented according to the obtained profiles and the defined SLAs that are based on fuzzy logic. Section 7.2 explains in detail the monitoring suite and the predictors used to detect the need for changes in many parameters obtained from machines. Section 7.3 concludes the chapter.

7.1. The dynamic SLA controller

This section presents the system that evaluates the need of updates on a virtual router environment. This system dynamically controls SLA and provides QoS guarantees for a virtualized network environment.

7.1.1. *Background on QoS for virtual networks*

The proposed dynamic allocation mechanisms found in the literature are mainly focused on server consolidation, and virtual router resource allocation is not well covered. Sandpiper [WOO 07] is a system that monitors virtual machines in a data center and migrates virtual machines to different physical servers in order to achieve a distributed virtual machine configuration that maximizes the performance and reduces the misuse of resources. Sandpiper analyzes the resource usage profiles of each virtual machine by using time series to avoid misbehaviors such as denial of service (DoS) attacks. The resource utilization of each physical server is define by a volume metric Vol, expressed as:

$$Vol = \frac{1}{1 - cpu} * \frac{1}{1 - mem} * \frac{1}{1 - net}, \quad\quad [7.1]$$

where cpu is the processor utilization percentage, mem means the memory utilization and net stands for network utilization.

Meng *et al.* [MEN 10] propose algorithms that provide the best allocation of virtualized servers in a grid of physical servers in order to improve network scalability and optimizing bandwidth utilization in the communication links. Virtual servers are instantiated on physical servers in order to reduce the distance between servers that are exchanging data with each other. Menascé *et al.* [MEN 06] apply autonomic computing techniques to control the processor sharing among virtual machines. The authors propose a dynamic allocation algorithm that is validated through simulations.

Xu *et al.* [XU 08] propose control mechanisms based on fuzzy logic to optimize the resource allocation in data centers and also execute performance tests in virtualized Web servers with different workloads. A learning system feeds the fuzzy controller to describe the Web server behavior under different loads.

One important tool that helps the dynamic resource allocation is the migration of virtual machines, which is used in Sandpiper [WOO 07]. The migration primitive enables migration of virtual machines from one physical server to another, allowing preventive maintenance and energy saving obtained through the reorganization of machines and shutdown of underutilized servers. Nevertheless, the virtual machine migration procedures, when applied on a virtual router application, represent a huge challenge due to the packet losses during the period of time when the machine is suspended. Wang *et al.* [WAN 07] propose a live migration mechanism without packet loss and Pisa *et al.* [PIS 10] implement this proposal in the Xen architecture.

The dynamic resource allocation and control is a challenge in the Xen platform because the input/output (I/O) virtualization technology is still naive, without isolation, allowing malicious virtual routers to impact the performance of other virtual routers in the same physical router. Therefore, XNetMon [FER 10a] proposes a secure control system for routing traffic based on data and control plane separation approach to manage the use of I/O resources by the virtual routers.

Keller *et al.* [KEL 09] analyze the QoS requirements in virtualized environments and propose an authoring model to account for and guarantee SLAs in virtual routers. The authors justify the authoring mechanism advantages and propose two main implementation models. The first model is based on network parameter monitoring while the second is based on the use of trusted platforms and cryptographic keys to provide tamper-proof properties.

This chapter presents a dynamic controller based on SLAs for virtual networks. The control is based on the generation and further analysis of router profiles, on the detection of SLA violations and on real-time punishment of misbehaving virtual routers. The proposed system monitors load values associated with each virtual network and generates real estimates of use profiles. These profiles ensure resource allocations that reduce the overload probability. The load function is based on the Sandpiper proposal but also includes other important parameters, such as system robustness, operation processor temperature and others that can be added if desired. The proposed fuzzy control system aims at providing an easy weighted parameter configuration. The fuzzy logic maps administrator strategies and detects SLA violating routers. Besides, the network administrator can easily insert new rules and action strategies. The proposed system supports data/control plane separation and, as a consequence, is fully compatible with XNetMon.

Results obtained from the prototype show the behavior of the control system, the generated profiles and the strategy mechanisms. The controller consumes only a few central processing unit (CPU) cycles to execute, allowing the monitoring of many virtual routers in parallel.

7.1.2. The proposed fuzzy control system

The proposed fuzzy control system monitors and guarantees SLAs in virtualized network environments. The key idea relies on the generation and analysis of use profiles of each virtual network, on the detection of SLA violations and on the real-time punishment of misbehaving virtual routers. The SLA violation detection procedure considers the severity of the violation and maps the severity onto a punishment level that is applied to each misbehaving virtual router. The system load also influences the punishment level. Thus, the global state of routers and control domain is also characterized by using a fuzzy controller that takes into account the processor, memory, network and robustness (existence of redundancy and fail-proof mechanisms) when calculating the system load. According to the output of the controller, it is possible to dynamically modify the punishment strategies and the system tolerance to violation actions.

The proposed system is a distributed management composed of controller agents. Each agent controls a set of physical routers and their associated virtual routers, as seen in Figure 7.1. Each physical router has a control domain where a control and monitoring daemon monitors the allocation of physical resources, verifies the conformity of SLAs in real time and generates the use profiles of each virtual router. Five modules compose the controller agents. The strategy and policy module (SPM) holds management strategies that can be applied over physical routers under its control and updates the current strategy to be applied on each control and monitoring daemon. The service level module (SLM) keeps a database that associates the SLAs with each virtual router. The Knowledge Base Module stores the history, usage profiles, and the description of violations that have occurred. This module estimates future network migrations, detects the need for updates and also renegotiates contracts, adapting them to the real profile of each virtual router. The Actuator Module executes within the control and monitoring daemons and retrieves the profiles and statistics of each virtual router. The controllers also have a Communication Module, which allows the exchange of information among other controllers through secure channels. If necessary, the controllers may use those channels to negotiate actuation domains' changes and negotiate virtual elements migration.

The described system has three main mechanisms. The first mechanism is the profile generation mechanism, which provides utilization statistics and

SLA violation detection. This information is stored in the KBM in order to renegotiate possible misconfigurations in the contracted SLAs. The second mechanism is the system load estimator, which gives an output that combines multiple resource status in an estimate within the $[0, 1]$ interval. The third mechanism is the adaptive punishment mechanism. Based on the system load and use profiles, the mechanism uses a fuzzy controller that outputs a level of punishment, proportional to the overall state of the system. For instance, if the system presents a low load, a medium violation (e.g., overcome in 20% the SLA) generates a small punishment (reduce in 2% the resource utilization of the machine that violates the SLA). On the other hand, if the system is overloaded, without available resources, even a small violation can be punished severely.

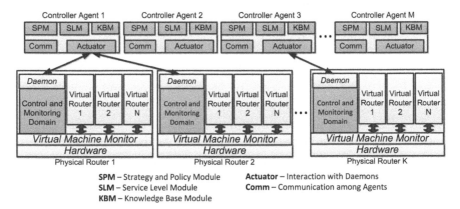

Figure 7.1. *The control system architecture. Actuator module of the distributed controller agents interacts with monitoring and control daemons running on physical routers*

7.1.2.1. *Generating router profiles*

The use profile of each virtual router represents the resource consumption pattern of each virtual router. The profiles are used to detect rule violation, estimate resource consumption and predict future needs. Profiles are generated through the capture of memory, processor and network utilization over the time to store the behavior of the recent past and long past of these variables. There are two distinct-size sliding windows to store the two associated time series. The generation of profiles based on probabilistic density functions is used in Sandpiper [WOO 07]. Recent past is used to check SLAs while the long past is used to predict future behaviors and estimate possible future demands. Behaviors are analyzed by probability functions.

Probability density function (PDF) of a processor utilization by a given virtual router executing RIPv2 is shown in Figure 7.2a). The exchange of control and data messages in this specific router has generated a processor usage of approximately 0.7% of CPU in 70% of the measurements during the long past window, which consists of the last 200 measurements for this scenario. Considering this PDF, we conclude that it is acceptable to aggregate a group of virtual routers with similar resource patterns and make them share the same physical core, without losing performance. The prototype also provides cumulative distribution functions (CDF) to verify SLAs. Flexible SLAs are possible and we can define, for instance, that a virtual router can use up to 0.7% of a given resource for a maximum of 80% of its execution time. Through the CDF of the recent past window, it is possible to determine that the router in Figure 7.2b) would fulfill the given SLA. It is important to mention that the generated distributions can be applied to any router from any machine. The given example just demonstrates which kind of distributions can be obtained.

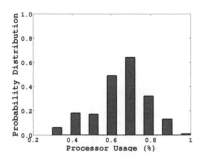
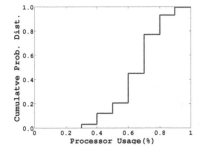

a) Probability density function (PDF)   b) Cumulative distribution function (CDF)

Figure 7.2. *Processor utilization profiles of a virtual router running RIPv2*

7.1.2.2. *Strategy and policy module*

The SPM stores the current acting strategies and maps administrative decisions into actions and strategies. We use fuzzy controllers [KEC 01] due to its capability to deal with decision-making problems that present uncertainties and qualitative parameters, for example the strategies of a network manager or administrator. In fuzzy logic, as mentioned before, a given element belongs to a given set according to its membership level inside the interval $[0, 1]$, where $\mu_A(x): X \rightarrow [0, 1]$ defines a membership function. We adopt the Mamdani Imply Method with Zadeh's [ZAD 00] AND and OR operators and the centroid method of defuzzification. The fuzzy controllers have small computational complexity and can parallelize inference procedures enhancing system performance and reducing the response time of the controller.

The SPM supports different acting strategies. Each strategy is composed of a set of inference rules, a set of membership functions that map input parameters according to the network manager perception (such as "high processor utilization" and "low memory load") and also a set of membership functions that regulates the output. There are two strategy types: the system load strategies and the punishment strategies. These two approaches and their correlated strategy package are described in section 7.1.2.5. Those strategies formalize a computational behavior that reflects the will and strategies of the network manager.

7.1.2.3. *Estimating the system load*

The system load is a measurement that determines the load level of the managed resources. Multiple parameters such as processor utilization, memory utilization, network utilization, system overall temperature and system robustness, which indicate the existence of redundancy mechanisms for disks and power supplies, can be analyzed to characterize the system state. We define the set of membership functions $\mu_{Proc}, \mu_{Mem}, \mu_{Net}, \mu_{Temp}$ and μ_{Rob}, which associates each of the resources in fuzzified variables. The combination of the parameters generates an output defined as system load, limited in the interval $[0, 1]$, which defines the load of the system. This value is used as an input parameter of another controller, together with delta, which is the difference between the resources contracted and the resources used, to estimate the punishment level of routers that violate SLAs. A block diagram that represents the system load controller is shown in Figure 7.3.

The resource use of each virtual router is aggregated to generate the controller input. An example of a possible configuration that evaluates the processor use and system temperature can be seen in Figure 7.4. It is important to remember that the presented curves can be modified according to each manager's needs and qualitative thoughts. Low, medium, high, cold, warm and hot are membership functions. In the given configuration, we have used three membership functions to map each resource. The definition of membership functions represents the mapping of the qualitative decision of the network manager. Most of the rules can be defined as triangular or trapezoidal function. For example, if the processor use is high and the overall system temperature is high, then the system overload is also high.

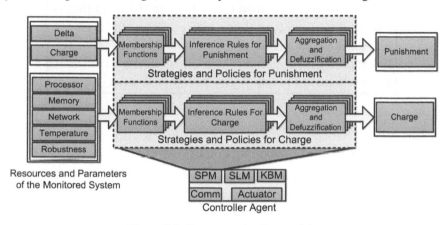

Figure 7.3. *Strategy and policy module*

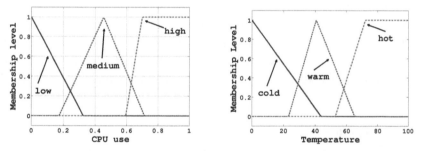

a) Membership functions for processor utilization

b) Membership functions for temperature

Figure 7.4. *Membership functions for processor and temperature*

7.1.2.4. *Strategies based on inference rules*

The fuzzy controller strategies are based on the default fuzzy inference rules of the fuzzy system. These rules follow the IF → THEN pattern, which represents the current action strategy scheme. The set of rules that defines as strategy is called a strategy package. An example of strategy package that calculates the punishment level according to the difference between the contracted SLA and the current resource use, denoted by delta, and the system load can be seen in Table 7.1.

Strategy packet
If **delta** (low) and **load** (low), then **punishment** (low)
If **delta** (average) and **load** (low), then **punishment** (low)
If **delta** (high) and **load** (low), then **punishment** (average)
If **delta** (low) and **load** (high), then **punishment** (average)
If **delta** (average) and **load** (high), then **punishment** (high)
If **delta** (high) and **load** (high), then **punishment** (high)

Table 7.1. *Example of a piece of a strategy package*

The presented strategy package corresponds to a network manager policy that establishes that even huge SLA violations are not punished severely, when the system is lightly loaded, because the system has plenty of resources and at this moment, giving additional resources to the violating router do not disturb the others. However, when the system is overloaded, the network manager is more rigorous and even light violations are punished severely. These rules work together with the membership functions that can be also developed by the network manager. New rules and strategies are easily inserted and the controller agent must export the strategy packages to the daemon that must use the defined strategy. We can also establish different strategies for each resource under control, enhancing the flexibility of the controller.

7.1.2.5. *Load policies*

The inference rules provide fuzzy values that represent the pertinent level of each inference rule and then these values are mapped onto a single controller output, which is a value in the interval $[0, 1]$ representing the current system load. Figure 7.5 shows two possible load policies, a conservative load policy and an aggressive load policy. Depending on the network manager profile, we can dynamically change the current domain policies.

Figure 7.6 shows decision surfaces of the system, which represent the punishment level for a system load and delta, that is the SLA violation level. The conservative policy only severely punishes a misbehaving virtual router when delta is high and the system is overloaded (Figure 7.6a)) while the aggressive policy generates high levels of punishment even for small positive variations in system load and delta (Figure 7.6b)).

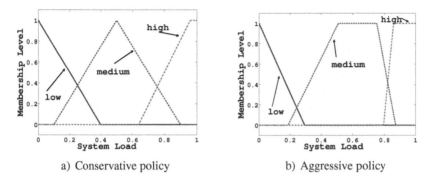

a) Conservative policy   b) Aggressive policy

Figure 7.5. *System charge policies*

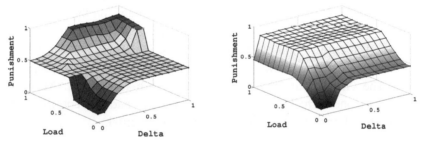

a) Decision surface for a conservative   b) Decision surface for an aggressive policy
policy

Figure 7.6. *Decision surfaces to different management strategies*

7.1.2.6. *Controlling the system overload and SLAs*

The developed system generates utilization profiles, evaluates if the profiles correspond to established SLAs, computes an estimate of the system load, and punishes virtual routers that violate the proposed rules. The daemon that executes within each domain performs the parameter gathering at each given sampling interval, which can be defined by the network manager. The parameters are used to generate temporal series that represent the utilization behavior of each resource and also the statistics and distributions that allow the verification of profiles and the compliance with SLAs. All this information is sent to the responsible controller agent. The daemon verifies if each virtual router profile corresponds to the negotiated SLAs. Moreover, it aggregates the resources used by each router to estimate the total load of the physical system. If a virtual router violates the contract, the system generates a value that represents delta. The system then uses this delta value and the system load to decide which is the appropriate punishment level to be applied on the misbehaving router. In the Xen architecture, we use the cap control parameter. The cap regulates the number of CPU cycles each virtual element can use. Thus, varying the cap value, we can control the use of processor resources of virtual routers. Another control tool that can be used is the Traffic Control (TC), which allows the queue control, permitting the management of the throughput of each virtual router, if they violate SLAs.

7.1.3. Results

The developed fuzzy control system is efficient and flexible. We have developed a set of experiments that prove the efficiency and the low overhead introduced by the proposed system. Tests were performed in a physical machine with a core i7 860 processor with four real cores and 8 GB DDR3 random access memory (RAM). The machine was configured with hypervisor Xen 4.0. The virtual routers were instantiated with 128 MB RAM memory and access to one virtual core. The virtual routers and the control domain execute Debian Lenny Linux with 2.6.32-5-amd64 kernel.

The design of the control system minimizes the processing overhead of the monitoring and control daemon. To evaluate the processor overhead of the daemon, we have instantiated some virtual routers and measured the average processor utilization of the control domain according to the number of virtual routers monitored. The results of Figure 7.7 show the processor utilization in the control domain according to the number of monitored virtual routers. In this configuration, measurements and decisions were gathered and evaluated at each second. The points in the graph represent the average processor utilization for each configuration with a confidence interval of 95%. We can see that the relation between processor utilization and number of monitored routers is approximately linear and even in situations where the daemon manages concurrently eight routers, the overhead load is acceptable and reaches up to 40% of processor utilization of a single core.

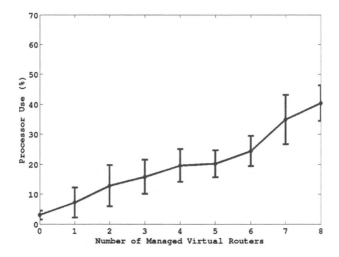

Figure 7.7. *Processor utilization in control domain when varying the number of managed virtual routers*

We can estimate that a control domain with the same configuration as the one used in this experiment can manage up to 20 virtual routers at the same time by dedicating only one single core to this task. Therefore, the system presents a good performance because it is monitoring multiple variables from multiple virtual routers at the same time and that the system is managing the SLAs of each virtual router.

The second experiment evaluates the controller efficiency and the effects of the punishment mechanism. We selected one of the virtual routers. The SLA of this selected virtual router defines that it can use up to 85% of processor of a single core to execute packet forwarding. Next, we create a packet flow that is forwarded by the router. When the flow is forwarded, the router violates the SLA and the control system regulates processor utilization through the cap mechanism. In the experiment, the punishment system is enabled when the router is already violating the SLA.

In the proposed scenario, we define three background environments. In the first enviroment, there is one monitored router and a virtual router that does not use resources; so, the system keeps a low load. In the second environment, there is the monitored router and five more virtual routers that are consuming a moderate amount of resources. In this scenario, the load was medium. The third enviroment has the monitored virtual router and seven extra virtual routers, both using almost all of the available resources. In this case, the system load is characterized as high. Hence, the characterization of the scenario comprises low, medium and high load outputs. These values were obtained considering a set of membership functions and inference rules that were on the system. In each configuration, the system generated a different load output depending on the background environment. We selected three specific environments from all the environments tested to demonstrate the different behaviors of the system for each possible system load output. Results shown in Figure 7.8 demonstrate that the system converges to ensure the router SLA. Depending on the load level of the system, the punishment level varies. We observe that in the low-load environment, the system takes up to 40 s to reach the contracted SLA. When there are plenty of resources, the SLA violation does not harm other routers. The use of the controller in the medium-load scenario incurs an increase of the punishment level and, as a consequence, in less than 15 s the abuse is contained. In the high-load scenario, the punishment is severe and the system limits resources within 5 s. Thus, the proposed controller is efficient and fulfills the established requirements, acting in a conservative way in low-load scenarios and acting aggressively in critical situations.

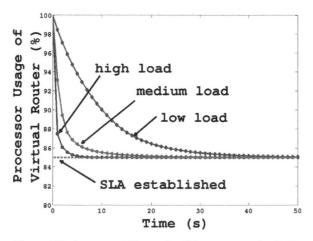

Figure 7.8. *System stability under different system loads. Faster convergence to the contracted SLA due to more rigorous punishment when the system load is high*

The third experiment evaluates the controller efficiency for transient misbehaving, when a given virtual router violates the SLAs for a given period of time and then it starts to comply with the contracts. The virtual router forwards packets at a high packet rate and uses 100% of its processor. After 60 s, the router forwards packets at a lower rate and consumes up to 80% of its processor, no longer violating the SLA. In this result, the system load is controlled and classified as medium. Figure 7.9 shows that without the controller, the virtual router may use whatever it wants to, possibly harming other routers. When the controller is active, the virtual router suffers a gradual caps reduction until the processing consumption of the virtual machine converges, complying with the SLA.

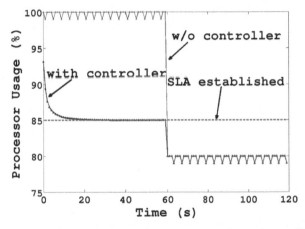

Figure 7.9. *processor utilization of a virtual router that violates a SLA for a given period, with medium system load*

7.2. Update prediction mechanism for local information

We propose an anomaly detection for autonomous management of virtual (ADAGA) networks system, that provides mechanisms for collecting and analyzing data in virtual network environments. ADAGA managers observe the monitored systems in the network, such as servers and routers, physical or virtual. The main objective of the proposed system is to send alarms to report possible anomalies occurring with the network entities. This study characterizes an anomalous behavior as short-term changes in the observations that are inconsistent with past measures.

ADAGA uses time series to predict the actual value considering the past of the series and to compare with the new observation. ADAGA considers that all time series initiate at zero. For that, the initial value of each series is decremented of all its observations' values that allow zero error during the initialization and no influence of the initial condition on future predictions. The correct predictor initialization is an important configuration, because it impacts the performance of the whole system [BRU 00].

The predictor analysis is based on the false positive and false negative values when predictor parameters are varied. This report analyzes the behavior of the monitored systems, describes anomalous situations and shows the impact on the emission of alarms when an anomalous situation occurs. The experiments are performed on a real router and the anomalies were artificially generated to simulate an overload in the router. The results show that ADAGA system detects anomalies presenting average false positive and false negative rates.

Section 7.2.1 discusses the studies related with the autonomous management and anomaly detection. Section 7.2.2 describes the developed system and its modules. Section 7.2.3 presents the testbed scenario of the experiments with the implemented prototype and the results obtained. Finally, section 7.3 concludes and presents directions for future work.

7.2.1. *Background on anomaly-detection systems*

Anomaly detection-based virtual network monitoring and management is not a well-described topic. Anomaly detection techniques are commonly used in security such as intrusion detection systems [PAT 07] and have also been used in autonomous systems for network management, which are triggered by the detection of an anomaly and the emission of an alarm. Anomalies can be classified into three types [BAR 01]: anomalies in network operation, which consist of network equipment failures or configuration changes; anomalies caused by a flash crowd, which usually happens when particular location information is requested by many users at the same time, such as the distribution of a new version of an operating system or a viral video; and anomalies caused by network abuse, such as DoS attacks and port scanning. The proposed system considers all of these anomalies because they are significant for the users' satisfaction of autonomous virtual network management.

Brutlag [BRU 00] uses time series and predictive mechanisms for anomaly detection in computer networks. The author focuses on the analysis of network traffic anomalies from a router to generate alarms. Results for single parameter configuration of the predictor are presented with initialization at zero, without any data preprocessing in series. In ADAGA, we preprocess the received data in order to obtain zero error at start up. Moreover, the key difference between ADAGA and Brutlag's system is the multidimensional analysis, where the calculations of time series are applied to different network characteristics such as memory and processor utilization, network traffic and system load. In virtualized scenarios, because of low network isolation, not only does the network traffic impact the network performance, but also processor and memory utilization. Thus, all of these ADAGA monitored metrics are important in virtualized scenarios [FER 10b]. Besides, the network management operations, such as virtual machine migrations, impact the network equipment operation [PIS 10].

De Lucena and Moura [DE 09] analyze network traffic focusing on packet-flow-based observations. The authors define flows by a 4-tuple (source IP address, destination IP address, source transport port and destination transport port). The flow approach defines various types of anomalies such as DoS, configuration failures and flash crowds. In the ADAGA system, the flows are not grouped in the conventional way. The system brings together the packages by services, because our purpose is to manage the network from the point of view of their function, considering that the pluralistic network approach considers one virtual network per service on the future Internet [MOR 09]. Therefore, in the ADAGA system, the group of packages take into account protocol number and destination transport port, because they are packet characteristics that define the flow service.

Several works related to anomaly detection present observation intervals of the order of units or tens of minutes [BRU 00, DE 09, SIL 10a, SIL 10b], which reduce processing and storage requirements. However, long-time intervals of this magnitude do not allow a quick reaction to important anomalies that happen in a short period of time. The ADAGA system offers observations spaced approximately by 15 s, and, therefore allows quick detection and reaction to anomalies. The experiments with the prototype present satisfactory results because the process of collection and analysis is accomplished in real time and the analysis of 40 different characteristics takes 10^{-4} s with an Intel Core i7 950 processor. Besides the sampling interval, another important characteristic of intrusion detection systems is the packet sampling rate. Per-packet processing systems require high processing and storage loads if they evaluate all packets. For this reason, several works perform packet sampling [SIL 10b, SOU 05, BAR 02]. Nevertheless, packet sampling introduces distortions, noise and smoothing on the observations [BRA 06]. Recent proposals address this problem through tagging potentially anomalous packets for further analysis in other network equipment [ALI 10] or through filters more efficient than random sampling [BRA 10]. Our proposal collects traffic statistics instead of packet sampling; therefore, all packets are considered with less processing overhead.

Following the chain of autonomous management processes described by Dobson *et al.* [DOB 06], that is collection, analysis, decision, and acting, the anomaly detection requires discovering the root cause of the anomaly. The root cause is obtained from the most recent observations of the network [SIL 10a, PAR 10]. ADAGA does not address discovering the root cause of the anomalies, but sends the latest observations with all collected data to be processed by other root cause discovering mechanisms.

7.2.2. ADAGA system

The ADAGA system provides mechanisms for collecting and analyzing data in a virtual network environment and supports the autonomous management of virtualized networks. The proposed system aims at detecting anomalies on virtual networks and also activating anomaly fixing mechanisms. Figure 7.10 shows the architecture of ADAGA, as well as the modules' interconnection and the communication among the monitored systems and manager.

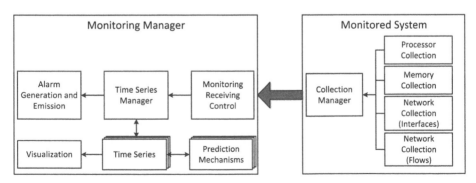

Figure 7.10. *ADAGA system architecture*

An ontology to represent and organize the knowledge of the network at any given moment is defined in section 7.2.2. The network administrator selects the relevant characteristics to be monitored and each of these characteristics is organized into a time series, as defined in section 7.2.2. Time series have an associated prediction mechanism, as described in section 7.2.2, which measures the deviation of the current measure in this characteristic past, generating alarms described in section 7.2.2. The virtualization module generates time series' graphic evolution of the predictor error, as described in section 7.2.3.

7.2.2.1. *Data collection and representation*

The ADAGA system collects data from network equipment through remote data requests on the *Extensible Markup Language* (XML) format. The monitoring manager enquires monitored systems that can be either physical or virtual elements in the network, as shown in Figure 7.10. Each monitored system executes the monitoring agent that, on receiving the requests, invokes several specialized agents to obtain specific data such as processor utilization memory utilization and network state.

The observation metric is fundamental in a monitored network scenario [ZIV 05]. Our prototype collects observations through the available tools of Linux operating system, which is the base system used in ADAGA, and of Xen [BAR 03], which is the chosen virtualization platform. Figure 7.11 shows a simplified representation structure for the network elements in ADAGA.

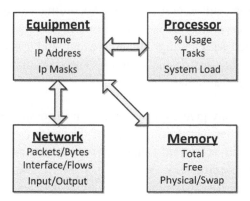

Figure 7.11. *Representation structure of network components in ADAGA system*

The network elements' modeling considers multi-core processors, RAM mechanisms, i.e. physical and virtual memory, and several network interfaces. The network equipment model includes identification and location data of equipment and connection with different processors, memory and network interface models. Each characteristic participates in the anomaly detection as distinct time series. In the ADAGA system, the calculations described in section 7.2.2 are applied for each characteristic. We claim that an effective anomaly detection system must be able to detect an anomaly by only analyzing the data statistics instead of the real collected data. This feature enables the proposed system to have a more effective processing and storage, in addition to avoiding problems with privacy of the data packets.

7.2.2.2. *Time series*

According to Brockwell and Davis [BRO 02], time series are a set of observations s_t, each one of them performed in a specific instant $t \in T$, where T is the finite group of the measurement times. The difference between a time series and a common set of values is that the order among the observations is important. There are two types of time series considering the observation process. In discrete time series, the observations have defined times and are performed in specific moments of time. In continuous time series, the observation is performed continuously during an interval of time. In this work, discrete time series are used.

We define the set T by the sequence of observation times. We also consider that the sampling interval between the observations follows the Poisson distribution with center in 15 s. According to Paxson [PAX 98], a fixed sampling interval may cause distortions in the observations because it can be synchronized with an unpredictable event, and it is unable to correctly observe periodic behaviors on the network. These problems reduce detection accuracy or hide the anomalies.

In time series, at any given specific time t, it is possible to predict the next value s_{t+1} of the series based on the past of the series $(s_1, s_2, \ldots, s_t)$. Based on this prediction and on the real value observed at the time $t+1$, it is possible to define the predictor error as:

$$\epsilon_t = \left| \hat{s}_{t+1} - s_{t+1} \right|. \hspace{3cm} [7.2]$$

If these errors are bigger than the tolerance defined in the prediction, ADAGA triggers the alert generation module, as described in section 7.2.2.

The time series management module feeds and controls the time series on the fly with the received observations. This module defines the monitored characteristics of each router, the predictor parameters of each characteristic and the insertion of new values on the characteristics time series.

7.2.2.3. Prediction mechanisms

Prediction mechanisms determine the next value of each time series. Two prediction mechanisms are considered: the simple and well-used predictor, named *exponential smoothing*, and *Holt-Winters seasonal*, which considers the trend and the seasonal components of time series. The *exponential smoothing* prediction mechanism is a simple algorithm to calculate the next value in a time series, which is based on the moving average of the past of the series [BRU 00]. To calculate each next value, $\hat{s}_{t+1}$, the expressions for the current measured value, s_t, and the prediction calculated for the current value, $\hat{s}_t$, are

$$\hat{s}_{t+1} = \alpha s_t + (1 - \alpha)\, \hat{s}_t, \tag{7.3}$$

where $\alpha \in [0, 1]$ and $\hat{s}_1 = s_1$. The α parameter is the weight of the current value in relation with the past of the series. Therefore, the bigger the α value, the smaller the influence the past of the series has on the calculation of the predicted value. According to the analysis of De Lucena and Moura [DE 09], on the computer networks scenario, appropriated values for the α parameter must be lower than 0.1. Because of the recursions performed as from $t \geq 2$, the influence of the past of the series is reduced throughout the time, following the expression given by:

$$\hat{s}_t = \left[\sum_{j=0}^{t-2} \alpha(1 - \alpha)^j s_{t-j} \right] + (1 - \alpha)^{t-1} s_1. \tag{7.4}$$

Therefore, Figure 7.12 shows that the influence of each past observation on the prediction calculation decreases exponentially, except by the first value. Thus, the initial condition strongly impacts the results, as observed in Figure 7.12a). Hence, to reduce the impact of the initial condition on the prediction of values, this work applies a conversion on the values of the series such that

$$\forall s_t, s_t = s_t - s_1. \qquad [7.5]$$

Then, the initial configuration will be $\hat{s}_1 = 0$ without prediction error, as shown in Figure 7.12b).

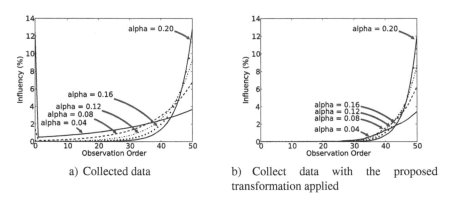

a) Collected data

b) Collect data with the proposed transformation applied

Figure 7.12. *Influence of the past observations in the prediction of the $t = 50$ observation for the transmitted packet time series*

The *exponential smoothing* mechanism is not suitable for time series presenting periodical behaviors, called seasonality, because it assumes linear series values and approximates the next value with the moving average of the series history. The *Holt-Winters seasonal* mechanism is a predictor well adapted to seasonal time series. Brutlag [BRU 00] defines a model of seasonality for the behavior of computer network, which consists of a greater activity during the morning than during the night.

The *Holt-Winters seasonal* decomposes the time series in tendency, seasonality and noise. Each one of these components is handled as a variation of the *exponential smoothing* method. There are two methods for aggregating these components into a predicted value: additive and multiplicative [KOL 99]. These components are aggregated with the additive method when the statistic variation of the period does not depend on the series. In case it does depend, the components are multiplied. De Lucena and Moura [DE 09] claim that, for computer networks, the additive composing of the components presents better results and, thus, the next value prediction is:

$$\hat{s}_{t+1} = R_t + T_t + P_{t+1-m},$$ [7.6]

where T_t represents the tendency of the time series, P_{t+1-m} is the periodic component, m is the seasonality period and R_T is the series aggregated noise. Then, the prediction equations are given by:

$$R_t = \alpha (s_t - P_{t-m}) + (1 - \alpha)(R_{t-1} + T_{t-1})$$ [7.7]

$$T_t = \beta (R_t - R_{t-1}) + (1 - \beta) T_{t-1}$$ [7.8]

$$P_t = \gamma (s_t - R_t) + (1 - \gamma) P_{t-m},$$ [7.9]

where α, β and $\gamma \in [0, 1]$ represent the smoothing constants for each component of the predicted value. Similarly to the *exponential smoothing* mechanism, these parameters represent the weight of the past series on the predicted value calculation. The bigger the constant values, the smaller the influence of the past component on the prediction.

7.2.2.4. *Alarm generation*

Alarm generation occurs when the predicted error is bigger than the calculated accepted error. The accepted error is recalculated for each new collected observation and is defined by:

$$\epsilon_t = \delta \Psi_t,$$ [7.10]

where δ is an amplification constant of the error acceptance and Ψ_t is defined according to the predictor mechanism. Brutlag [BRU 00] claims that optimum values of δ belong to the interval $[2, 3]$. ADAGA uses $\delta = 2$, because we propose a sensitive system and $\delta = 2$ generates a smaller acceptance error.

For the *exponential smoothing* mechanism, Ψ_t is the standard deviation of the values considering the observation window and the next value prediction. For the *Holt-Winters seasonal* mechanism, Ψ_t is

$$\Psi_t^{HOLT} = \gamma(|s_t - \hat{s}_t|) + (1 - \gamma)(\Psi_{t-m}^{HOLT}).$$ [7.11]

The ADAGA system proposes an alarm control with the cumulative emission of alarms. Then, not all generated alarms are emitted to remove punctual alarms of the system. The implementation of this method in ADAGA uses a hysteresis to define the emissions of generated alarms. If the system detects an anomaly during the observations, this alarm is not emitted immediately. Only after generating η alarms, one alarm is emitted. The value of η is a counter threshold defined by the network administrator. If the system stops detecting anomalies before achieving η value, the anomaly is not reported and the alarm accumulation counter is set to zero.

An alarm emission means sending a report to the decision system and acting on the network. This report consists of a group of the latest 15 observations of all characteristics of the network element, which generated the alarm, the information of predictor value the real observed value and the characteristic that generated the alarm.

7.2.3. *The anomaly system evaluation*

The ADAGA system was evaluated to determine its capacity to detect anomalies. The evaluation considers false positive and false negative rates. False positive means that alarms were wrongly emitted because there were no anomalies, and false negatives are characterized by observations at moments with anomalies that do not generate alarms, therefore, including the alarms accumulated during the period of alarm generation hysteresis.

We developed a prototype to analyze the ADAGA system with the testbed scenario shown in Figure 7.13. The monitoring manager executes the time series management, predictions and alarms generation. The monitored system is a wireless network router used in a real network. The collected information is periodic with a collecting interval following the Poisson distribution with a center in 15 s [PAX 98].

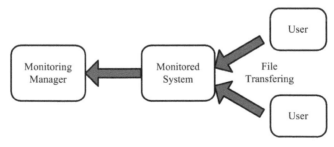

Figure 7.13. *Testbed scenario for analysis of false positives and false negatives in ADAGA*

After, approximately, 2 days of monitoring, two nodes started two consecutive uploads of a 15 GB file to the router using the *Secure Shell* (SSH) protocol to overload the router and avoid remote access. The objective is to cause an anomaly that must be detected by ADAGA and, as a result, send the alarms. Besides, we evaluate the impact of each characteristic on the detection of this anomaly. As expected, time series of network characteristics, such as network traffic in the reception interface and the flow at Transmission Control Protocol (TCP) port 22, detect the anomaly, whereas other characteristics, for example number of running processes and memory utilization, do not. We observe that when the system load variable presents a particular behavior, such as a lot of peaks, an anamoly has been detected. The system load is influenced by the file transfer because the performed upload induces network and disk I/O load in the router.

7.2.3.1. *Results*

The series of observations of the TCP packet reception in SSH service are shown in Figure 7.14. Figure 7.14a) shows the results for the *exponential smoothing* mechanism and Figure 7.14b) shows for the *Holt-Winters seasonal* mechanism. At the top of the graphs, there are the real observations, the solid line, and the mechanism-predicted values, the dashed line. From this graph, we observe that the *exponential smoothing* mechanism-predicted values easily follow the real-value evolution, which makes anomaly detection more difficult, even for anomalies of this magnitude. Differently, the *Holt-Winters seasonal* mechanism has greater difficulty in following the observed values, showing better anomaly detection efficiency. When a predictor is able to follow abrupt changes, it will have low efficiency for detecting anomalies, which are defined as abrupt changes. The bottom part of the graphs shows the values directly connected with the anomaly detection. The solid line represents the prediction error of the mechanism, obtained according to equation [7.2] and the dashed line shows the accepted prediction error that is calculated by equation [7.10]. For best visualization, the graphs only show the time interval when the anamoly has occurred.

In Figure 7.15, we observe a past behavior of two monitored characteristics; the average load of the monitored system in 5 min before each data collection, in Figure 7.15a), and the percentage of the processor utilization, in Figure 7.15b). From a graphical analysis, we observe the characteristics, such as the system load, that suffer from the influence of network traffic anomalies. This influence confirms the importance of the multidimensional network equipment monitoring. Anomaly signals of different characteristics are useful for tracing the cause of anomalies. Characteristics such as the processor utilization are not influenced by the generated anomaly as shown in Figure 7.15b).

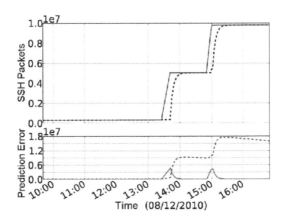

a) *Exponential smoothing mechanism*

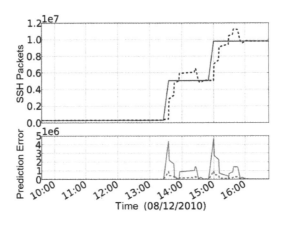

b) *Holt-Winters seasonal mechanism*

Figure 7.14. *Temporal evolution of TCP packet receiving in SSH service for each mechanism. Observation window of 2,000 samples and α, β and γ are equal to 0.1*

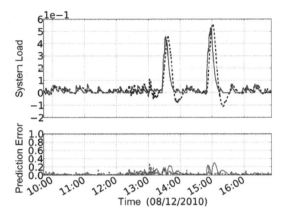

a) Average load of the system in 5 min before each observation

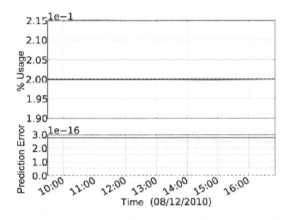

b) Percentage of processor utilization on the router

Figure 7.15. *Temporal evolution of two characteristics not directly related to network traffic for the Holt-Winters seasonal mechanism. Observation window of 2,000 samples and α, β and γ are equal to 0.1*

The false positive rate computes the number of emitted alarms when there are no anomalies, for all the collected observations. Moreover, the false negative rate is the number of alarms not issued during anomalies for all collected observations.

As expected, the *Holt-Winters seasonal* mechanism presents better results than the *exponential smoothing* mechanism as shown in Figure 7.14.

Figure 7.16 shows the percentage of false positive and false negative for several parameter configurations of the prediction mechanisms as we vary η parameter, which represents the quantity of alarms accumulated before the report emission. The *exponential smoothing* is evaluated for α equal to 0.05, 0.10 and 0.15. The same values were applied to α, β and γ parameters of the *Holt-Winters seasonal* mechanism, resulting in the 27 curves shown in Figures 7.16c) and (d). All graphs in Figure 7.16 are considered for an observation window of 2,000 samples. The tests performed for other sizes of observation window present similar results in the induced anomaly detection.

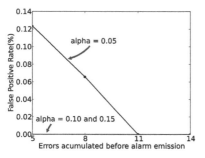

a) False positive rate of *exponential smoothing* mechanism

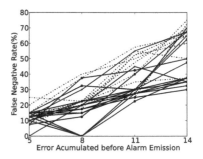

b) False negative rate of *exponential smoothing* mechanism

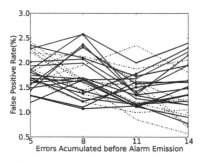

c) False positive rate of *Holt-Winters seasonal* mechanism

d) False negative rate of *Holt-Winters seasonal* mechanism

Figure 7.16. *Analysis for the TCP packet receiving in SSH service characteristics of each mechanism, when α, β and γ parameters change. Observation window size is 2,000 samples*

Although the *exponential smoothing* mechanism presents low false positive rates as shown in Figure 7.16a), it is not able to detect well-generated anomalies. On the other hand, the *Holt-Winters seasonal* mechanism presents a good effectiveness for all the evaluated parameter configurations, with false positive rates lower than 2.5%, and false negative rates appropriated with the defined accumulation of alarms. During the anomaly interval, we collected 40 samples. Table 7.2 presents the false positive rates, and Table 7.3, the false negative rates for all the evaluated parameter configurations of the *Holt-Winters seasonal* mechanism. The rows are the false positive and negative values for each value of α, and the columns present the values of false positive and negative for each value of γ, grouped by the values of β. We can conclude that there is a big similarity between the results for different configurations. Nevertheless, we observe a minor gain for smaller β values. The β parameter is related with the series tendency; therefore, the more the predictor accepts drastic changes on the time series tendency, that is for great values of β, the faster the predictor is able to adapt itself to the anomaly, and as a result, the smaller the rates of success are.

β	0.05			0.10			0.15		
$\alpha\backslash\gamma$	0.05	0.10	0.15	0.05	0.10	0.15	0.05	0.10	0.15
0.05	1.69	1.34	2.19	1.64	1.49	1.81	1.44	1.85	1.59
0.10	1.33	1.19	2.20	2.35	1.85	2.28	1.55	1.93	2.42
0.15	1.80	1.59	1.65	2.28	1.74	1.88	2.01	1.68	2.32

Table 7.2. *Comparative table of false positive rates for the α, β and γ configurations shown in Figure 7.16c) with $\eta = 5$. Values are expressed in percentage*

β	0.05			0.10			0.15		
$\alpha\backslash\gamma$	0.05	0.10	0.15	0.05	0.10	0.15	0.05	0.10	0.15
0.05	15.0	12.5	7.5	10.0	15.0	12.5	7.5	15.0	10.0
0.10	12.5	15.0	25.0	0	12.5	15.0	15.0	10.0	15.0
0.15	7.5	12.5	15.0	12.5	10.0	15.0	15.0	10.0	20.0

Table 7.3. *Comparative table of false negative rates for the α, β and γ configurations shown in Figure 7.16c) with $\eta = 5$. Values are expressed in percentage*

7.3. Summary

Our pluralist approach consists of dedicated virtual networks to provide different specialized or general-purpose isolated virtual networks over a unique hardware that provides different QoSs. We also propose a piloting plane that must manage and control all the virtual networks for handling all the problems arising in the network such as: congestion, failure and QoS provision. The piloting plane uses information provided by the knowledge plane to autonomously execute management and control procedures. Managing and controling the whole network is a huge challenge, and we claim situated view solutions to help these activities. Situated view is a partial view of the whole network limited to their neighborhood and the surroundings. In this chapter, we have presented two manage and control systems.

The first proposed system is based on the development of a dynamic resource allocation system that analyzes the resource utilization profiles of virtual routers and provides a fair share of resources based on QoS metrics and SLA agreements. We have developed an efficient fuzzy controller for SLA control of virtualized network environments, in which isolation prerogatives represent a huge challenge. The obtained results show that the proposed system is efficient and fits well for network resource control. Network managers may easily insert rules that reflect their personal and qualitative decision-making strategies. The obtained results show that the system efficiently controls the SLAs by punishing misbehaving routers that violate rules. Furthermore, the punishment level depends on the system load and the level of violation. In our experiments, the system has successfully limited the SLAs adaptively. When the network is a state with plenty of idle resources, the system applies light punishments and at critical moments the system applies severe punishments. The results also show that the controller adjusts the resource use in less than 5 s. In a low-load condition, the system converges within 40 s. Besides, the monitoring and management generate only a small overhead in the control domain that corresponds to 5% of a single processor for each managed virtual router.

The second proposed system focuses on monitoring and prediction techniques that monitor the environment, provide proper and updated information to the knowledge plane and also detect router misbehavior. We have developed ADAGA networks system that collects and analyzes data from virtual networks. The major advantages of ADAGA are multidimensional monitoring, low monitoring interval, which provides low reaction times, and avoidance of per-packet processing strategies. The proposed system considers time series and minimizes the overhead by considering counters and statistics instead of data packets. The results show that the *exponential smoothing* mechanism is not suitable for computer network predictions, because these networks present seasonality. The false positive and false negative results show that the *Holt-Winters seasonal* mechanism is efficient, because it detects the network anomalies with low false positive rates and with false negative rates.

The two systems presented in this chapter help users and agents to manage and control virtual networks and also to be aware of the need for updates. The first system approach monitors and manages the resource utilization profile and SLAs to maintain QoS in networks and to estimate the need of SLA updates to obtain each virtual router's demands. By generating the resource utilization profiles, it is possible to detect violations and guarantee SLA constraints. The second system monitors several parameters and its use predicts future behaviors, to detect possible changes and, then, to trigger alarms to force the system to keep the information refreshed in the knowledge plane.

7.4. Bibliography

[ALI 10] ALI S., HAQ I., RIZVI S., *et al.*, "On mitigating sampling-induced accuracy loss in traffic anomaly detection systems", *ACM SIGCOMM Computer Communication Review*, vol. 40, no. 3, pp. 4–16, 2010.

[BAR 01] BARFORD P., PLONKA D., "Characteristics of network traffic flow anomalies", *Proceedings of the 1st ACM SIGCOMM Workshop on Internet Measurement*, ACM, pp. 69–73, 2001.

[BAR 02] BARFORD P., KLINE J., PLONKA D., *et al.*, "A signal analysis of network traffic anomalies", *Proceedings of the 2nd ACM SIGCOMM Workshop on Internet Measurement*, ACM, pp. 71–82, 2002.

[BAR 03] BARHAM P., DRAGOVIC B., FRASER K., *et al.*, "Xen and the art of virtualization", *Proceedings of the 19th ACM Symposium on Operating Systems Principles*, ACM, pp. 164–177, 2003.

[BRA 06] BRAUCKHOFF D., TELLENBACH B., WAGNER A., *et al.*, "Impact of packet sampling on anomaly detection metrics", *Proceedings of the 6th ACM SIGCOMM Conference on Internet Measurement*, ACM, pp. 159–164, 2006.

[BRA 10] BRAUCKHOFF D., SALAMATIAN K., MAY M., "A signal processing view on packet sampling and anomaly detection", *INFOCOM, 2010 Proceedings IEEE*, IEEE, pp. 1–9, 2010.

[BRO 02] BROCKWELL P., DAVIS R., *Introduction to Time Series and Forecasting*, Springer Verlag, 2002.

[BRU 00] BRUTLAG J., "Aberrant behavior detection in time series for network monitoring", *Proceedings of the 14th USENIX Conference on System Administration*, New Orleans, LA, USA, 3–8 December 2000.

[DE 09] DE LUCENA S., DE MOURA A., "Análise dos Estimadores EWMA e Holt-Winters para Detecção de Anomalias em Tráfego IP a partir de Medidas de Entropia", *CSBC2009*, 2009.

[DOB 06] DOBSON S., DENAZIS S., FERNÁNDEZ A., *et al.*, "A survey of autonomic communications", *ACM Transactions on Autonomous and Adaptive Systems (TAAS)*, vol. 1, no. 2, pp. 223–259, 2006.

[FER 10a] FERNANDES N.C., DUARTE O.C.M.B., "XNetMon: Uma Arquitetura com Segurança para redes virtuais", *Anais do X Simpósio Brasileiro em Segurança da Informação de Sistemas Computacionais - SBSeg10* Fortaleza, CE, Brazil, pp. 339–352, October 2010.

[FER 10b] FERNANDES N., MOREIRA M., MORAES I., *et al.*, "Virtual networks: isolation, performance, and trends", *Annals of Telecommunications*, vol. 66, no. 5–6, pp. 339–355, 2011.

[KEC 01] KECMAN V., *Learning and Soft Computing: Support Vector Machines, Neural Networks, and Fuzzy Logic Models*, MIT Press, 2001.

[KEL 09] KELLER E., LEE R., REXFORD J., "Accountability in hosted virtual networks", *Proceedings of the 1st ACM Workshop on Virtualized Infrastructure Systems and Architectures*, ACM, pp. 29–36, 2009.

[KOL 99] KOEHLER A., SNYDER R., ORD J., "Forecasting models and prediction intervals for the multiplicative holt-winters method", *Monash Econometrics and Business Statistics Working Papers*, 1999.

[MEN 06] MENASCÉ D., BENNANI M., "Autonomic virtualized environments", *2006 International Conference on Autonomic and Autonomous Systems*, ICAS 06, , IEEE, p. 28, 2006.

[MEN 10] MENG X., PAPPAS V., ZHANG L., "Improving the scalability of data center networks with traffic-aware virtual machine placement", *INFOCOM, 2010 Proceedings IEEE*, IEEE, pp. 1–9, 2010.

[MOR 09] MOREIRA M., FERNANDES N., COSTA L., *et al.*, "Internet do futuro: Um novo horizonte," *Minicursos do Simpósio Brasileiro de Redes de Computadores-SBRC* 2009, pp. 1–59, 2009.

[PAR 10] PAREDES-OLIVA I., DIMITROPOULOS X., MOLINA M., *et al.*, "Automating root-cause analysis of network anomalies using frequent itemset mining", *ACM SIGCOMM Computer Communication Review*, vol. 40, no. 4, pp. 467–468, 2010.

[PAT 07] PATCHA A., PARK J., "An overview of anomaly detection techniques: existing solutions and latest technological trends", *Computer Networks*, vol. 51, no. 12, pp. 3448–3470, 2007.

[PAX 98] PAXSON V., "On calibrating measurements of packet transit times", *SIGMETRICS/PERFORMANCE: Joint International Conference on Measurement and Modeling of Computer Systems, Madison*, WI, 1998.

[PIS 10] PISA P., FERNANDES N., CARVALHO H., *et al.*, "OpenFlow and Xen-Based virtual network migration", *The World Computer Congress 2010 – Network of the Future Conference*, Brisbane, Australia, pp. 170–181, September 2010.

[SIL 10a] SILVEIRA F., DIOT C., "URCA: pulling out anomalies by their root causes", *INFOCOM, 2010 Proceedings IEEE*, IEEE, pp. 1–9, 2010.

[SIL 10b] SILVEIRA F., DIOT C., TAFT N., *et al.*, "ASTUTE: detecting a different class of traffic anomalies", *Proceedings of the ACM SIGCOMM 2010 Conference*, ACM, New Delhi, India, September 2010.

[SOU 05] SOULE A., SALAMATIAN K., TAFT N., "Combining filtering and statistical methods for anomaly detection", *Proceedings of the 5th ACM SIGCOMM Conference on Internet Measurement*, USENIX Association, p. 31, 2005.

[WAN 07] WANG Y., VAN DER MERWE J., REXFORD J., "VROOM: virtual routers on the move", *Proceedings of the ACM SIGCOMM Workshop on Hot Topics in Networking*, Citeseer, 2007.

[WOO 07] WOOD T., SHENOY P., VENKATARAMANI A., *et al.*, "Black-box and gray-box strategies for virtual machine migration", *Proceedings of Networked Systems Design and Implementation*, Cambridge, MA, USA, April 2007.

[XU 08] XU J., ZHAO M., FORTES J., *et al.*, "Autonomic resource management in virtualized data centers using fuzzy logic-based approaches", *Cluster Computing*, vol. 11, no. 3, pp. 213–227, 2008.

[ZAD 00] ZADEH L., "Fuzzy sets*", *Information and Control*, vol. 8, no. 3, pp. 338–353, 1965.

[ZIV 05] ZIVIANI A., DUARTE O., "Metrologia na Internet", *Minicursos do XXIII Simpósio Brasileiro de Redes de Computadores*, SBRC, pp. 285–329, 2005.

Chapter 8

System Architecture Design

The future Internet should support multiple networks simultaneously, each with its own protocol stack and management schemes, assuring the network core great flexibility. Accordingly, this chapter describes an architecture to enable the creation of a programmable network core. The aim is to design virtualized infrastructures using current virtualization technologies. For that, three different technologies are used: the Xen hypervisor, the OpenFlow switch architecture and a combination of these two, which integrates machine and network virtualization techniques.

The use of both Xen and OpenFlow platforms capitalizes on the advantages of these two machine and network virtualization platforms. OpenFlow networks give a wide support for flow migration and virtual network remapping without packet losses, while Xen provides a more flexible packet processing due to the use of virtual machines. The idea is to use OpenFlow for managing flows and Xen for providing control of the network and routing packets. Such an approach is called XenFlow and it achieves a flexible and complete set of network management tools.

This chapter provides a general view of the developed modules, which allows the creation of a coherent prototype that facilitates the development of new networks' address-specific issues, such as the provisioning of mobility,

Chapter written by Otto Carlos M.B. DUARTE.

security and Quality-of-Service (QoS). The integration of the developed tools in both Xen and OpenFlow platforms is defined in this chapter as well as how the designed modules interact with each other in an efficient fashion.

Moreover, this chapter specifies the piloting system and the service control requirements for designing policy-based architectures. Specifically, it develops a network management framework for post-IP networks that correlates and optimizes each network element to compose a global network with self-management capability. This chapter also defines a thin management layer within virtualized substrates that is autonomic, capable of situated awareness, learning, inferring and detecting faults for adaptive monitoring in order to provide self-properties to the system. Besides that, a piloting system is introduced and the integration of separate self-control functions is discussed.

The distributed architecture for the management of virtual networks and a piloting system prototype enable the network substrate to self-manage virtual networks. The autonomic managers of network elements have a closed control loop of monitoring, analyzing, planning and executing; such a loop feeds information to the knowledge base for next iterations of the loop. The testbed, used to test this concept, is composed of virtual machines working as routers over a physical network infrastructure. A prototype of our architecture using a multi-agent system based on the Ginkgo platform was also developed. The testing scenarios focus on self-healing of virtual networks, but the distributed architecture for self-management of virtual networks is sufficiently generic and it can be used for other functionalities in autonomic computing such as self-configuration, self-optimization and self-protection. Some experiments are presented to show the performance of the recovery process.

This chapter is organized as follows. Section 8.1 provides the overall architecture view of both Xen and OpenFlow platforms. It provides a detailed view of how the proposed algorithms interact with each other as well as how virtualization management tools are used by control algorithms. Auxiliary functions, such as plane separation and secure communication, are also described. Section 8.2 presents the XenFlow architecture design, which is a hybrid architecture that combines Xen and OpenFlow virtualization platforms; a performance analysis of the main features of this new platform is also shown. Finally, section 8.3 concludes the chapter.

8.1. Overall architecture design

In this section, the architecture of the developed prototype using Xen and OpenFlow virtualization platforms is presented. We show which algorithms are related to each platform and how tools interact. Some modules that were designed to create a consistent set of tools are also described in this section.

8.1.1. *The Xen architecture*

In network virtualization, different virtual routers share a physical router in order to provide different network services, simultaneously. Key aspects of this paradigm are isolation and performance on packet forwarding. Isolation ensures independent virtual network operation, preventing malicious or fault virtual routers interference in the operation of other virtual networks. Xen [FER 11] is a virtualization platform that can be used to create virtual routers, each with its own operating system and protocol stack, in commodity computers. The Xen platform, however, does not provide complete isolation and also presents low performance on handling network input/output (I/O) operations [EGI 07].

An architecture design is presented for monitoring and ensuring isolation, security and high performance in each physical node used by virtual networks (Figures 8.1 and 8.2). Figure 8.1 shows the entities' relationship in the proposed architecture for Xen virtual networks, whereas Figure 8.2 shows the software architecture in each physical node. The decision was made for the adoption of a modular architecture to guarantee an easy upgrade of a tool without disturbing other services. Moreover, it is easy to add new features to the proposed model, since the basis for the developed modules is the virtual network management for the Xen-based testbed tool (VNEXT) [PIS 11], which aggregates all the proposed control modules.

The architecture of the node is based on a client-server model, in which virtual machines are clients of a server located in Domain 0 (dom0). Available services are optional and can be selected according to the administrator requirements. Moreover, some services run only in dom0, and do not interfere with others in the virtual machine.

Most of the client-server functions used in our prototype are based on the plane separation paradigm [WAN 08], in which data forwarding and control mechanisms are decoupled. Thus, control, such as routing, is accomplished in the virtual machine (user domain (domU)), ensuring flexibility in the design of control mechanisms, while packet forwarding is performed in a privileged domain, called dom0, providing quasi-native performance. The main advantage of plane separation is the provision of quasi-native packet forwarding performance. A drawback, however, is the fact that the infrastructure administrator imposes a common forwarding plane for all virtual machines. Instead, we offer the virtual network operator the option of using, or not, the plane separation paradigm, which is responsible for creating a valid copy of the forwarding table of the virtual machine in dom0. The QoS operator module creates a copy of the forwarding rules of the virtual machine, such as queuing disciplines and filtering rules, inside dom0. Hence, these modules accomplish the transfer of the data plane from the virtual machine to dom0. Both modules use the secure channel module, to ensure secure communication between domU and dom0.

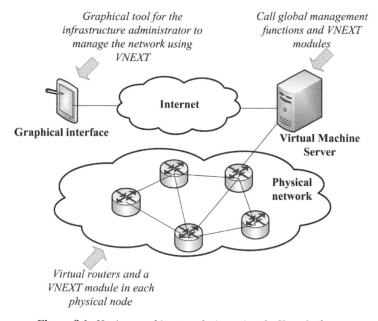

Figure 8.1. *Horizon architecture design using the Xen platform*

The monitoring module is also based on a client-server model and provides a set of monitoring data for dom0 and for each virtual machine. Besides that, the modules for migration [PIS 10], topology discovery and virtual network configuration are described. All these modules compose the basic tool set for controlling network virtualization.[1]

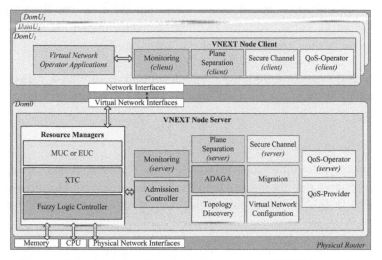

Figure 8.2. *A physical node using the Xen platform: VNEXT server in Domain 0 and VNEXT clients in Domain U*

The anomaly detection for autonomous management of virtual networks system (ADAGA) module detects anomalies both in the physical and in virtual nodes. ADAGA interacts with the monitoring module to obtain data about domUs and dom0 and performs actions for creating filtering rules to monitor the virtual networks.

The main modules of the architecture are the resource managers, which dynamically control resource allocation to the virtual machines. Three modules (maximum usage controller/efficient usage controller (MUC/EUC), Xen throughout control (XTC) and Fuzzy controller) were developed and an admission controller module was also created, which operates in parallel with the resource managers to evaluate whether or not new virtual networks can be hosted in the same physical node. This module gives support to global virtual network allocation functions.

Two different modules were added to the node architecture to provide QoS: the QoS-operator and the QoS-provider. The QoS-operator allows the virtual

1 The VNEXT tool evolved to Future Internet Tesbed with Security (FITS) and can be obtained at http://www.gta.ufrj.br/fits/

network operator to set QoS parameters in spite of the use of plane separation. The QoS-provider differentiates the service provided by each virtual network, allowing the infrastructure administrator to set different privileges for each virtual network. These modules facilitate the development of QoS support, which is one of the current requirements of the Internet.

In the following, a description of each module is provided, especially the admission controller and the QoS modules as well as how they interact with each other.

8.1.1.1. *Resource managers*

Our prototype assumes the use of up to three resource managers, depending on the resources that the infrastructure administrator wants to monitor. Each manager defines a different policy and controls a different parameter in the virtualized system.

The first resource manager module selects a resource policy, according to the selected controller. There are two options: the MUC [FER 11a] and the EUC. MUC reserves a minimum amount of resources for each domU and shares the remaining resources among all domUs. There is no bound of resource usage for the virtual networks. Therefore, this policy allows a virtual network to use more resources than those specified in the service level agreement (SLA). The sharing of the remaining resources is a function of the value of a parameter set by the infrastructure administrator to differentiate virtual networks. The EUC specifies a minimum resource reservation rate and a maximum volume of resources to be provided in a long time interval to each virtual network. The controller dynamically adjusts the resource reservation parameters of each virtual network according to the reservation parameters and the network demand. Hence, EUC specifies the maximum amount of resources each virtual network can use, providing a more precise resource reservation scheme provided by MUC. Both the MUC and the EUC are correlated to the monitoring module, which provides the required information to perform control of resource allocation. The resource manager also interacts with the virtual network configuration to obtain and modify the resource allocation parameters of the virtual networks. The virtual network configuration module is initially set by the VNEXT virtual machine server (VMS), which exports the parameters selected by the infrastructure provider during network creation or management of each physical node, guaranteeing an accurate behavior in the controller modules.

The VMS allows the infrastructure administrator to set the resource parameters of either MUC or EUC to each virtual node by using a graphical interface. The default configuration assumes the same parameters for all nodes, but it also allows per-node configuration.

The second resource manager module is XTC [COU 11]. XTC is a mechanism for controlling the bit rate at which a virtual machine forwards packets, reducing its impact on dom0. Indeed, XTC dynamically adjusts Xen scheduler parameters to give more or less central processing unit (CPU) time to each virtual network in order to increase or reduce its packet forwarding rate. As XTC limits the bit rate that a virtual router sends, XTC is only effective when the plane separation of VNEXT architecture is not used. Like all VNEXT modules, the actions performed by XTC are carried out by the VMS. For this, each physical machine executes a daemon, called XTC Manager that can be connected to the VMS by Transmission Control Protocol (TCP) sockets. The VMS is thus responsible for receiving user's requests and taking the appropriate action by exchanging messages with XTC Manager. The XTC actions performed by the VMS can be done by using the graphical user interface (GUI) of VNEXT. This module provides a user-friendly interface to XTC by making requests through the VMS.

The third resource manager is a fuzzy logic controller [CAR 12]. This controller presents a complimentary functionality to the EUC and MUC, allowing the infrastructure provider to control other parameters rather than the shared CPU, memory, and bandwidth in dom0, such as the robustness to failures and machine temperature. The idea is to provide an efficient control system for SLAs in virtual network environments. The proposed system verifies the physical resource usage, retrieves real-time profiles of virtual routers and guarantees SLA requirements. The control is based on fuzzy logic and it determines the resource allocation according to the system overload and the profile of routers. The control logic punishes virtual networks that exceed the established SLA.

This controller also uses the monitoring module and generates profiles, represented by histograms, that contain relevant system information.

8.1.1.2. *Admission control of virtual routers*

The virtual network admission controller arbitrates the access of new virtual routers to the physical machine. The number of virtual routers hosted in any physical machine influences the long-term capacity of processing reservations. Indeed, EUC defines two resource reservation schemes:

– The *short-term rate reservation* ($R_s[n]$), which is the rate of resource reservation that must be met for the virtual network n in the short-term I_s.

– The *long-term volume of reservation* ($V_l[n]$), which is the amount of resources $V_l[n] = R_l[n] \cdot I_l$ that should be guaranteed during a long-term interval I_l, where $R_l[n]$ is the long-term average rate volume reservation request for the virtual network n.

Indeed, the admission control mechanism is non-trivial since different requests may require the same long-term resources. The profiles of demands, which include static information, such as disk and memory sizes, and dynamic information, such as throughput and CPU consumption, should be considered in the calculation of the probability of a virtual network service request which is to be denied, in order to prevent overloading physical resources. Therefore, the proposed admission control stores a set of histograms representing the physical-substrate resource usage [FER 12]. A new histogram is generated for each resource at the end of a monitoring period. A set of histograms of the same resource models the load variation for a period of time (e.g. a day). Histograms represent the aggregate resource usage by virtual networks. Networks with different traffic profiles lead to different aggregate resource histograms, even if the short- and long-term reservations are the same for all virtual networks. Two networks can present the same V_l and R_c and show completely different behaviors, which imply different aggregated resource demands.

A monitoring interval is defined as $K_{\mathrm{adm}} \cdot I_l$ seconds, where K_{adm} is an arbitrary constant chosen by the infrastructure administrator. To avoid storage and processing overload, the proposed system randomly selects K_{rand}, which is the number of long intervals that will not be evaluated after K_{adm} long intervals. Moreover, instead of storing all histograms, the admission controller checks the difference between the current and the last histogram. For this, the admission controller normalizes both histograms and calculates the longest error in the y-axis (e_m) between the two histograms. If condition $|e_m| > E_{\mathrm{adm}}$ holds, where E_{adm} is a threshold specified by the infrastructure administrator, then the current profile is stored and a new histogram is started.

The controller algorithm estimates the blocking probability after a new virtual network joins the physical substrate. This estimated probability is used as a criterion for the decision on the acceptance of a new virtual network. The inputs of the algorithm are the substrate histograms, the long-term volume reservation of each virtual network ($V_l[\ ] = R_l[\ ] \cdot I_l$) and the average resource usage of each virtual network, $R_{avg}[\ ]$.

The admission control algorithm is accomplished in four steps:

– *Step 1:* estimate the aggregate resource usage considering a histogram.

– *Step 2:* estimate how the increase in the demand impacts the aggregate resource usage (assuming that the reservations are respected).

– *Step 3:* estimate a function to model the new virtual network resource usage.

– *Step 4:* calculate the resource blocking probability if the new network was accepted.

Step 1

Step 1 consists of monitoring resource usage and storing histograms as described previously. At the end of this step, the algorithm knows the substrate histograms for bandwidth, CPU and shared memory of each monitoring interval.

Step 2

In this step, the proposed algorithm estimates the substrate histograms based on the hypothesis that all virtual networks were using all reserved resources. The idea is to ensure that there will always be physical resources to meet all the virtual networks' requirements, in despite of simultaneous peak demands. Solutions based on migration [WOO 09] are usually slow, because they depend on observing the resource blocking for a period of time, then searching for a new mapping between the virtual and physical topologies, to avoid overloading physical nodes, and, finally, migrating the selected virtual nodes. This solution can cause packet losses, leading to penalties for the infrastructure provider. Migration-based solutions without the use of an appropriate admission control can overload the physical nodes, causing losses and oscillations in the mapping of virtual networks on the physical substrate.

Therefore, it is important to estimate the impact of the admission of a new virtual network on the substrate before admitting it into a physical node. The key idea of our proposal is to estimate resource blocking using the hypothesis that the new virtual network request was accepted and all virtual networks were fully utilizing the associated physical resources.

The proposed algorithm estimates the demand increase as:

$$\Delta = \sum_{n=1}^{N_H} R_l[n] - R_{\text{avg}}[n], \qquad [8.1]$$

where N_H is the number of virtual networks. The variable Δ estimates the impact of the increase in the demand on the physical substrate. For this estimation, it is assumed that virtual networks will request all reserved resources and that the demand does not increase as a function of the order of reservation requests. The function $f_i(t)$ models the resource usage of the virtual network i and an estimation of the increased resource usage of this network is given by $f_i(t) + \Delta_i$, where $\Delta_i = R_l[i] - R_{\text{avg}}[i]$. Nevertheless, it is irrelevant for the proposed algorithm the way each network increases its consumption individually, but the way the aggregated usage increases when each virtual network uses its whole reservation. After obtaining Δ (equation [8.1]), the controller updates the histogram. First, the histogram is shifted according to Δ. $I[i]$ is a histogram interval with upper bound $L_s[i]$, and $I[i']$ is an interval that contains the value $L_s[i] + \Delta$, then

$$L_s[i' - 1] < L_s[i] + \Delta \leq L_s[i']. \qquad [8.2]$$

Furthermore, $H_{\text{sub}}(I[i])$ is the number of events occurrences in the interval $I[i]$ and N_{int} is the number of intervals in the histogram. Thus, the shifted histogram, H_{sft}, is computed as $H_{\text{sft}}(I[i']) = H_{\text{sub}}(I[i])$. The interval $I[N_{\text{int}}]$ of H_{sft} is defined as:

$$L_s[N_{\text{int}} - 1] < I[N_{\text{int}}] < \infty, \qquad [8.3]$$

to maintain a fixed number of intervals. Figure 8.3 shows an example of histogram shift when $N_{\text{int}} = 10$ and $\Delta = 1$. After this step, the controller calculates a probability mass function (*PMF*) for the shifted substrate histogram (*PMF_{hist}*), assuming that the values of the x-axis are given by $L_s[s]$, as shown in Figure 8.4.

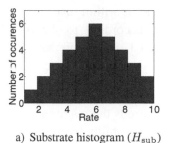

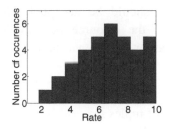

a) Substrate histogram (H_{sub})

b) Shifted substrate histogram (H_{sft})

Figure 8.3. *Example of substrate histogram shifting, when $N_{\mathrm{int}} = 10$ and $\Delta = 1$*

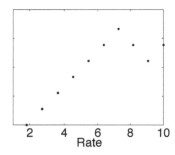

Figure 8.4. *Probability mass function of the shifted substrate histogram shown in Figure 8.3b)*

Step 3

The controller estimates the demand of the new virtual network based on a predefined distribution. For example, an optimistic admission controller assumes a Poisson distribution, while a pessimistic admission controller assumes a distribution concentrated at peak rate values.

The distribution estimated for the new virtual network (PMF_{new}) is represented according to the intervals $I[n]$ of the substrate histogram, assuming that the values of x-axis are given by $L_s[\,]$.

Step 4

The controller algorithm estimates the blocking probability for each resource. This probability is used to decide whether or not a new virtual network should be accepted. The proposed algorithm uses PDF_{hist} to calculate the resource blocking probability, according to the distribution assumed for the new virtual network (PDF_{new}). Considering that

$$A = \{(x_1, x_2) \mid x_1 + x_2 \geq C\}, \quad x_1, x_2 \in [L[0], L[N_{int}]], \qquad [8.4]$$

is a set of tuples (x_1, x_2) in which $x_1 + x_2$ exceeds the resource capacity C, the controller computes the resource blocking probability by using

$$P_B = \sum_{\forall(x_1, x_2) \in A} PDF_{hist}(x_1) \cdot PDF_{new}(x_2). \qquad [8.5]$$

Therefore, the proposed algorithm estimates the probability of the sum of the resources used in the shifted substrate histogram and if the new network exceeds the physical capacity.

The new virtual router is accepted if there are enough resources for the static requirements and if the condition

$$\sum_{n=1}^{N} R_s[n] < C \quad \text{and} P_B < P_L \qquad [8.6]$$

holds for all histograms of all resources, where P_L is the value of the resource blocking probability bound established by the infrastructure administrator. A small P_L value guarantees a low probability value of packet losses. Nevertheless, it also reduces the efficiency of resource utilization and, as a result, benefits the infrastructure provider.

8.1.1.3. *Plane separation modules*

As mentioned earlier, for the achievement of high performance, it is essential to forward packets via dom0. However, by separating the packet forwarding from the routing control, the virtual routers are unable to update their forwarding tables and their packet filters, because they have no access to dom0 memory region. Our system monitors the content of tables and filters in domUs and makes a replica of them in dom0 in the data plane manager module. Therefore, the client in each domU monitors changes in the forwarding table and packet filters, and the server in dom0 maps both the forwarding table and the packet filter built in each domU onto dom0.

Every change in the forwarding table or packet filter in domU must be immediately updated in its replica in dom0. For this reason, after every control message arrival, the data plane manager client checks for changes in the forwarding table and packet filters in domU. If any difference is detected, the client transmits the changes to dom0 via the secure communication module. When the data plane manager server receives a message notifying a forwarding table change, it searches for the settings of that domU to find out in which dom0 table to insert the change. In packet filter updates, the server modifies the received rule by inserting rule parameters that specify the characteristics of the virtual network. This procedure avoids one virtual router from creating rules in the packet filter that influence the traffic of other virtual routers.

8.1.1.4. *Secure communication module*

The secure communication module creates secure communication channels between the dom0 and the domUs, providing mutual authentication and privacy in data transfer. A secure communication channel is required since the data plane aggregates all forwarding tables of all virtual networks. Privacy and authentication are mandatory for guaranteeing isolation. Since this module is often used to update the data plane in dom0, it must be light. Mutual authentication is required to ensure that no domain can forge the identity of a common domain or the identity of dom0 to generate misleading information in the data plane that corresponds to the attacked domain.

The secure communication module is composed of two protocols: one based on asymmetric cryptography for exchanging session keys (K_s), as described in Figure 8.5a), and the other based on symmetric cryptography for a secure transmission of data between domU and dom0, as described in Figure 8.5b), where kp is the private key, Kp is the public key, $E([M], key)$ is the encryption of the message M with the key for symmetric or asymmetric cryptography, $Sign([M], k)$ represents message M and its signature with k and id is the source node identity. These protocols avoid replay attacks, in which the opponent domain repeats old control messages to spoil information in the data plane of the attacked domain, using sequence numbers and nonces, which are randomly chosen numbers that should be used just once in the control messages. The proposed mechanism also changes the session key every time a sequence number reaches a maximum value to increase the robustness against replay attacks. It can be assured that the communication between dom0 and domU is secure since the system checks the authenticity, the privacy, and data non-reproducibility.

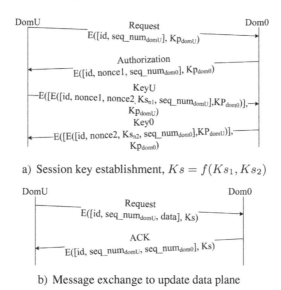

a) Session key establishment, $Ks = f(Ks_1, Ks_2)$

b) Message exchange to update data plane

Figure 8.5. *Message sequence for creating a secure channel between dom0 and DomU*

8.1.1.5. *QoS provision*

One important advantage of the proposed architecture is the support of QoS provisioning. Although a virtual router that does not use plane separation can use a traffic control mechanism control, it cannot ensure QoS. A virtual router has no control over its own traffic while forwarded by dom0 from the physical device driver to the virtual device driver. The I/O operations in Xen architecture are accomplished in dom0, which is the driver domain and, as a consequence, DomUs do not have complete control of the forwarding operations. When plane separation is adopted, the data plane is placed in dom0 allowing a fine control of the forwarding functions as well as QoS support. The proposed architecture offers primitives for adding QoS rules in virtual networks and among themselves, as shown in Figures 8.6a) and b) for packet forwarding through the virtual machine and through dom0, respectively. Figure 8.6a) describes in detail the QoS modules of the proposed architecture assuming packet forwarding through the virtual machine. According to this scheme, the infrastructure provider can configure priority access for the physical hardware to some virtual networks. Moreover, any virtual network operator can differentiate its own packets being processed inside the virtual machine.

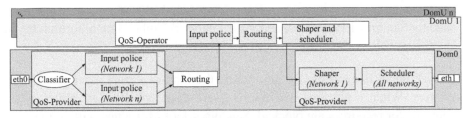

a) QoS provision assuming packet forwarding through the virtual machine

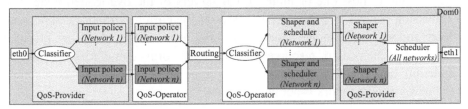

b) QoS provision assuming plane separation paradigm, in which the packet forwarding is exclusively performed in Dom0

Figure 8.6. *QoS provision inside each virtual network (QoS-operator) and among virtual networks (QoS-provider) in proposed architecture*

The use of plane separation implies that the data plane of each virtual network, which was previously in domU, is placed now into dom0. Thus, QoS provision must be changed to work with plane separation. As a result, QoS provision for both virtual network operator and the infrastructure provider must be supplied in dom0, as shown in Figure 8.6b). The proposed architecture provides an interface that allows virtual routers to configure their own QoS rules through the QoS-operator client and the QoS-operator server components. These components guarantee that one virtual network can configure its QoS rules without interfering with the QoS rules of other virtual networks.

8.1.1.6. *Analysis*

In this section, we analyze both the access control and QoS modules when used with the EUC controller. These modules were introduced to guarantee the support to QoS requirements.

8.1.1.6.1. Admission control module evaluation

The virtual network admission control for the proposed architecture was implemented in C++. The proposed algorithm was evaluated under different traffic patterns using simulation. The present proposal was compared to others, namely Sandpiper [WOO 09] and virtual network embedding algorithm based on the ant colony metaheuristic (VNE-AC) [FAJ 11].

Sandpiper is a system for monitoring the load of virtual machines in data centers. When a physical machine is overloaded, which is identified when requests for resources are blocked, Sandpiper looks for a new virtual machine that is able to host one or more virtual servers of the overloaded node. The admission control in Sandpiper is based on the current virtual machine peak load and on the average amount of idle resources in the new physical machine. If the peak of resource demand is 95% of its cumulative distribution function, or if it is less or equal to the average amount of idle resources, then the new virtual network is admitted.

VNE-AC is a virtual network mapping mechanism, which is part of the Horizon project, for the control of the global virtual network. This algorithm is based on the ant colony metaheuristic, which determines an adequate mapping of virtual networks over the physical substrate. The admission control mechanism proposed in the VNE-AC assumes that resources are statically attributed to each virtual network. As a result, this admission control restricts SLA specifications.

We implemented both Sandpiper and VNE-AC admission control to compare them to the proposed mechanism. The Sandpiper peak rate was chosen as $p_k = 95\%$, as suggested by its authors. VNE-AC reservation threshold was set to R_l, which is the average rate of the long-term volume reservation (V_l) in EUC. The impact of Δ (equation [8.1]), indicated as "Hist" on the results, was also evaluated to verify the efficiency of this procedure.

In the evaluation, the number of virtual networks that each mechanism admitted to be hosted in a physical router is measured. For simplicity, it is assumed that all virtual networks presented the same resource reservation parameters and that the number of intervals in the histogram for our proposal was $N_{\text{int}} = 30$.

The ideal number of virtual networks in the analyzed physical node in each experiment was also measured. This ideal number is chosen as the maximum number of virtual networks that guarantees that the blocking probability threshold specified by the infrastructure administrator (P_L) is not violated. Since $p_k = 95\%$ for Sandpiper, we selected $P_L = 0.05$ for a fair composition. On the one hand, if an admission control mechanism admits fewer networks than the ideal number, physical machine resources are wasted. On the other hand, if the admission control mechanism admits more virtual networks than the ideal number, physical nodes are overloaded that lead to penalties for the infrastructure provider. Admitting fewer virtual networks is preferable to admitting a number of virtual networks higher than the ideal value.

A total of 30 rounds for each experiment were executed and the output link throughput in packets per second was analyzed; the standard deviation of metrics of interest will be shown.

Traffic pattern impact: the first experiment evaluates virtual networks with traffic modeled by Poisson processes. Each virtual network demands $\approx$ 100 Mb/s, which is also the value of R_l. Figure 8.7a) shows that all mechanisms admitted the ideal number of virtual networks. Hence, if virtual network traffic presents a small deviation, then all the analyzed mechanisms are able to correctly perform admission control. Figure 8.7b) shows that admission control in this scenario is abrupt: 10 networks cause no blocking, while 11 networks cause 100% blocking.

To evaluate the proposals in environments with larger traffic variability, virtual networks were also simulated with traffic pattern described by an on-off model. Results are shown in Figure 8.8. In this experiment, all virtual networks present the same traffic pattern. The on-off traffic is generated based on an exponential distribution with $\mu = 1/3$. Time intervals are exponentially distributed. The traffic in the on state is modeled by a Poisson distribution with $\lambda \approx$ 200 Mb/s. Figure 8.8a) shows that Sandpiper and VNE-AC admit more virtual networks than the other proposals. Nevertheless, both mechanisms admit more virtual networks than the ideal number of virtual networks (Figure 8.8b)). Sandpiper underestimated the resources occupancy and allows the admission of approximately 10 networks. VNE-AC behaves similarly to Sandpiper. VNE-AC does not differentiate between the scenarios in Figures 8.7 and 8.8. Virtual networks with different data patterns are equally treated, generating misleading information to admission control. Our proposal admits a number of networks close to the ideal number, generating a low blocking probability and an efficient resource usage.

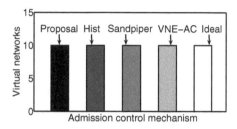

a) Number of virtual network admitted by each mechanism

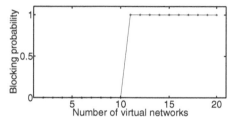

b) Blocking probability according to the number of virtual networks in parallel

Figure 8.7. *Admission control assuming virtual networks with traffic modeled by a Poisson process and maximum blocking probability of 5%*

Δ *Impact:*   This experiment evaluates virtual networks with varying demands. In this scenario, the throughput of the virtual networks increases up to the threshold of the long-term volume reservation. Admission request of new virtual networks is sent when the other virtual networks are half reserved. Figure 8.9a) shows that the algorithms Hist and Sandpiper accepted all the networks, since they do not consider the demand increase. Both our proposal and the VNE-AC consider the variable demands and allows a number of accepted virtual networks close to the ideal number.

Based on these results, it can be seen that the proposed admission control is the only analyzed mechanism that is efficient in all analyzed scenarios, guaranteeing the admission of a high number of virtual networks, without violating the blocking probability threshold imposed by the infrastructure administrator.

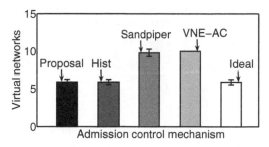

a) Number of virtual network admitted by each mechanism

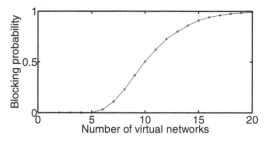

b) Blocking probability according to the number of virtual networks in parallel

Figure 8.8. *Admission control assuming virtual networks with on-off traffic and maximum blocking probability of 5%*

8.1.1.6.2. Evaluation of quality of service provision

The QoS components of the proposed architecture are evaluated [FER 12]. Figure 8.10 shows the effect of the use of QoS premises by the infrastructure provider. In this scenario, which is composed of two virtual networks, it is assumed that in Network 1 a small forwarding delay is faced by its traffic. The following parameter values are assumed for the output link: $R_s[1] = 50$ Mb/s and $R_l[1] = 400$ Mb/s for Network 1, and $R_s[2] = 100$ Mb/s and $R_l[2] = 600$ Mb/s for Network 2. CPU and memory resources are equally divided between the networks. Network input demands are, respectively, $D[1] = 50$ Mb/s and $D[2] = 1$ Gb/s. We chose a high value for $D[2]$, because a high volume of traffic hinders the provision of QoS to Network 1. The traffic of Network 2, which uses plane separation, flows between two external machines. Network 1 traffic, which is routed through the virtual machine, is generated by a different external machine and it shares the output link with the Network 2 traffic.

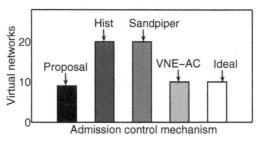

a) Number of virtual network admitted by each mechanism

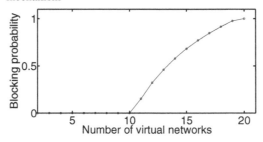

b) Blocking probability according to the number of virtual networks in parallel

Figure 8.9. *Admission control assuming virtual networks with increasing traffic and maximum blocking probability of 5%*

Figure 8.10 shows the round trip time (RTT) values in networks with and without the proposed QoS support. On the first scenario, labeled "w/o prio", both networks have the same priority and, in the second scenario, labeled "prio", the Network 1 traffic is privileged. When we give priority to one virtual network, even without the use of the resource sharing manager EUC, the RTT values decrease by more than 10 times for the privileged traffic. The use of the QoS support ensures that Network 2 exceeds neither the use of network nor the use of CPU, reducing the volume of the processed data and thereby reducing the transmission delays. Hence, the QoS module reduces the RTT value by more than 18 times when compared to the scenario without our proposal and without priority, and more than 1.8 times when compared to the scenario with privileged traffic but without the resource sharing manager EUC.

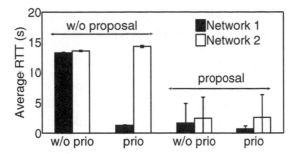

Figure 8.10. *RTT according to the QoS parameters inter virtual networks, assuming that Network 1 is prioritized*

8.1.2. *OpenFlow management architecture*

OpenFlow has an architecture different from Xen. In OpenFlow networks, the control plane is centralized in one node, while the data plane is distributed over the physical OpenFlow switches.

We developed the tool OpenFlow management infrastructure (OMNI), which is used as the basis for integrating OpenFlow management modules [MAT 11]. The overall architecture overview is shown in Figure 8.11.

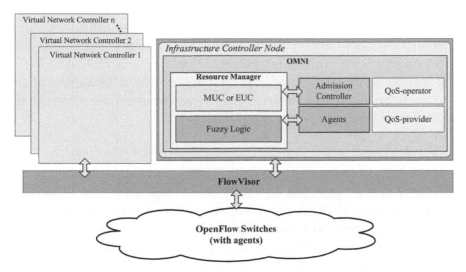

Figure 8.11. *Horizon architecture design using the OpenFlow platform*

Due to the centralized nature of the data plane, the proposed architecture for OpenFlow networks is simple. No modification is required in the forwarding nodes. Shared nodes do not notice their virtualization. Hence, all proposed functions are restricted to the controllers and the FlowVisor.

The original version of OMNI was designed to work with any OpenFlow controller, providing network view and intravirtual network management functions. The extended version of OMNI works coupled with FlowVisor to provide management functions to the infrastructure administrator. This extended version runs algorithms for controlling physical resources that are shared by all virtual networks.

The new OMNI modules are described on the right hand side of Figure 8.11. In this figure, we observe that all controllers interact with FlowVisor, which, in turn, interacts with the OpenFlow network. The OpenFlow network is composed of OpenFlow switches. Optionally, we can use agents in these nodes to increase management performance as well as to add new monitoring functions. Agents could also be aggregated in middle boxes or even in the infrastructure controller node to facilitate the performance of distributed control functions.

FlowVisor is an interface provided by the OpenFlow team to share the OpenFlow switches among several controllers [SHE 09]. FlowVisor offers an interface to configure physical switch resources, such as memory and queues. It does not, however, provide control algorithms. We applied the same controller algorithms used in Xen to control the resources in OpenFlow, since these algorithms are platform independent. Hence, we can apply the logic of the modules EUC, MUC, fuzzy logic, admission control, QoS-operator and QoS-provider in OMNI to control FlowVisor resources. We also add the agents to emulate a distributed control behavior.

8.1.2.1. *Evaluation of the OpenFlow management architecture*

The response time and operation of OMNI were evaluated and the number of control packets generated by OMNI was compared to that of network operating system (NOX) in order to estimate control overhead generated by OMNI. An experimental network was configured using personal computers running OpenvSwtich software [WAN 08], an implementation of OpenFlow. OpenvSwitch works as a Kernel module and assures high-performance packet forwarding. Our experimental scenario consists of four OpenFlow switches, a FlowVisor entity and a NOX controller, as shown in Figure 8.12. OpenFlow switches and the FlowVisor run on Intel Core 2 Duo computers with 2 GB of memory. The controller runs on an Intel I7 computer with 4 GB of memory. On this computer, Ginkgo agents, agents for controlling each OpenFlow switch, were also run. Confidence intervals with 95% confidence levels were desired.

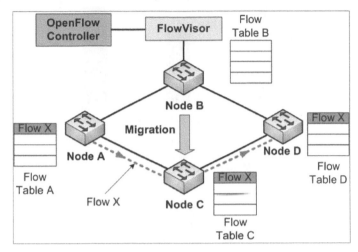

Figure 8.12. *Migration of the Flow x from path A-B-D to path A-C-D*

The first experiment evaluates the migration performed by the multi-agent system. This experiment consists of the migration of a flow from the path composed of switches A, B, and D to the path composed of switches A, C, and D (Figure 8.12). The probing traffic is a User Datagram Protocol (UDP) flow with 1,470 bytes of packet size and the rate varies from 0.5 to 3 Mb/s. In this scenario, the throughput of the AB link is upper bounded by OpenFlow at 200 kb/s, while the other links (BD, AC and CD) use their full capacity (Mb/s). Thus, packets loser occur when the transmission rate exceeds 200 kb/s. Packet losses are measured to autonomously trigger the flow migration. To decide on flow migration, the agent verifies the packet loss rate of its monitored switch, and also compares the local packet loss rate with the loss rates exchanged with other agents. The packet loss rate and the thresholds for comparison are provided to the agent as parameters and are set according to each network. The developed agent senses the network at fixed intervals of 10 s and migrates a flow only after three consecutive samplings indicate that the packet loss rate is above the threshold. Figure 8.13a) shows that the agent properly detects the bottleneck link within the flow path and then migrates the flow to avoid, or reduce, the packet loss rate. Since agents observe the loss rate in links at fixed time intervals, the time between starting the agents and the decision on flow migration is independent of the packet transmission rate. Monitoring indicates that the agent triggers the migration on average 29.4 s after starting the measurement procedure.

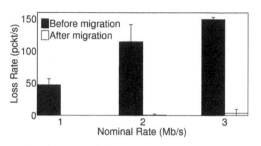

a) Packet losses before/after running the agents

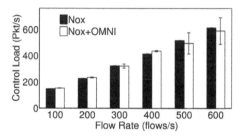

b) NOX and OMNI control load

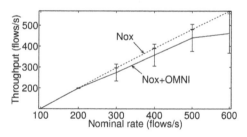

c) NOX and OMNI flow instantiation rate

Figure 8.13. *Results of migration experiment calling migration function by an agent and comparison of control overload between NOX and OMNI*

Flow instantiation is one of the main causes of control overhead in an OpenFlow network, because when a packet does not match any flow in a switch, the packet is forwarded to the controller and the controller sends a command to the switch. Thus, our next experiments evaluate the control overhead introduced by OMNI, and the effect of OMNI applications on flow instantiation. The rate of control packets that traverse an OpenFlow network is measured while new flows are instantiated using either NOX or NOX+OMNI. "NOX" designates a NOX controller running its original applications for collecting network statistics and for configuring packet forwarding while "NOX+OMNI" means running all OMNI applications on the NOX controller. Since the NOX+OMNI flow instantiation mechanism is the same as that of NOX, the difference between the two curves is due to the statistic monitoring of NOX+OMNI. Figure 8.13b) compares the control overhead for a varying flow instantiation rate. The difference between the two systems is negligible for instantiating up to 400 flows/s, since NOX and NOX+OMNI show almost the same control load. When instantiating more than 500 flows/s, the control load of NOX+OMNI is lower than that of NOX, but Figure 8.13c) shows that NOX+OMNI is unable to instantiate as many flows as NOX. The flow instantiation rate achieved with OMNI is greater than the rate of NOX+OMNI. Also, the error bar size of NOX+OMNI increases because of the test instability. Indeed, NOX+OMNI is trying to process more data than the controller is able to handle. Since OMNI interval for monitoring each resource is configurable, increasing this interval reduces OMNI overhead, which can lead to higher flow instantiation rate at the cost of increasing the granularity of statistic measures.

8.2. A hybrid Xen and OpenFlow system architecture design

Xen and OpenFlow have advantages and disadvantages. While OpenFlow presents simple flow configuration mechanisms, Xen is more flexible in packet processing. This has motivated the definition of a hybrid architecture to capitalize on the advantages of each platform, obtaining a powerful virtualization platform. The management tools [MAT 11, PIS 11] with the few modifications described in the previous sections also apply to this architecture.

The use of new protocols and services in the core of the Internet goes against the trend of most service providers, due to the high risk these changes represent for proper network operation as well as the high costs involved in changing the hardware of the platforms. Proposals for network virtualization to allow innovation in production networks are presented in [FEA 07, RAT 05], allowing isolated networks to share the same physical substrate.

Network virtualization introduces a new management primitive: the migration of virtual networks [WAN 08]. The migration primitive is used in different contexts, such as maintenance of physical nodes and remapping the logical topology over the physical topology. Maintenance of physical nodes often requires shutting down the device. In case of routers, the maintenance period causes adjacency losses and, consequently, network failures during the convergence of the routing algorithm. The virtual network remapping [ALK 12, ALK 11] is used for traffic management and for energy saving, in which virtual nodes are reorganized considering the energy demand [BOL 09, ROD 12]. Migration can also be used to prevent damages, for instance, under a denial of service (DoS) attack. In this scenario, virtual networks that share the same physical substrate with the network under attack are migrated to other nodes, preventing the overload in the input links. The migration of virtual topologies, however, presents great challenges, such as the relocation of virtual links and the decrease in the negative effects caused by the service downtime during migration.

There are proposals [WAN 07, PIS 10] that perform the migration of the logical topology in a transparent way to the network edges, without packet losses or broken connections. The scenarios in which these proposals are valid, however, are limited. In [WAN 07] and [PIS 10], it is assumed that a mechanism for link migration exists that is independent of the mechanism of node migration. It also assumes that one virtual router can only be migrated from one physical machine to another in the same local area network (LAN). Otherwise, tunnels between physical machines should be created to simulate a LAN.

However, flow migration in the OpenFlow platform is easy. Pisa *et al.* present an algorithm that is based on the definition of a flow path in the OpenFlow network [PIS 10]. This proposal presents zero packet loss and low control overhead. Nevertheless, OpenFlow migration is not applicable to router virtualization or flow processing systems. The proposal is limited to switched networks.

This section introduces the XenFlow architecture, which is a hybrid network virtualization platform based on Xen and OpenFlow. Our proposal describes a flow processing system that allows migration of virtual networks, including the migration of both nodes and links. In this architecture, the plane separation paradigm is used and then the virtual router is divided into two planes, the control and the data planes. The control plane, which runs in a Xen virtual machine, is responsible for updating the routing table, given the routing protocol decisions. The data plane, which is implemented using OpenFlow, is responsible for forwarding the packets according to routing policies. The routing policies are based on defined routes, calculated by the control plane. The main advantages of this network virtualization technique are as follows:

- Plane separation with a highly flexible data plane.

- Migration without packet losses.

- Migration not restricted to local network.

- Mapping of logical links onto one or more physical links.

- OpenFlow data plane, using distributed network control.

- Node and link migration are carried out using the same procedure.

A XenFlow prototype was built to validate the system architecture design. Experimental results show that the system carries out efficient migration, in the sense that there is no packet loss or routing service interruption during migration; the system allows migration of virtual routers and links without connection loss or packet forwarding delay. When comparing XenFlow migration to Xen virtual machine native migration, XenFlow showed zero packet loss, while Xen native migration looses a significant amount of packets and presents a longer downtime during control plane update.

8.2.1. *Pros and cons of Xen and OpenFlow virtualization platforms*

OpenFlow is a switching technology that enables programming packet forwarding by associating actions with flows. A flow is defined by a set of up to 12 fields extracted from the frame header, including data from link layer, network layer and transport layer headers [MCK 08]. The forwarding table of an OpenFlow switch is the Flow Table. The Flow Table relates a flow with one or more output ports of the switch according to the output actions defined by a centralized controller. The controller processes the first packet of a flow and, then, defines the actions. The Nox [GUD 08] is an OpenFlow controller that acts as an interface between the control applications and the OpenFlow network. As soon as a packet arrives at an OpenFlow switch, the switch checks if the packet matches any already defined flow. If so, the actions defined for that flow are applied to the packet. If not, the packet header is sent to the controller, which extracts the flow characteristics from the packet and creates a new flow in the Flow Table of the OpenFlow switch. An example of an OpenFlow network is shown in Figure 8.14. In OpenFlow networks, migration is carried out easily since it is performed by only reprogramming the Flow Tables in the switches.

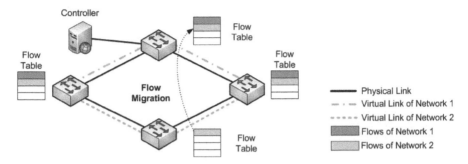

Figure 8.14. *OpenFlow network virtualization. Flow migration procedure is the redefinition of the network flows in another set of switches*

The two main disadvantages of OpenFlow networks are the need for a centralized control plane and the cost of per-hop packet processing.

Xen is a personal computer virtualization platform largely used for server consolidation. Its architecture is based on a virtualization layer, called virtual machine monitor (VMM) or hypervisor. The Xen virtual environments are called virtual machines, or domains, and present their own resources, such as, CPU, memory, disk and network access. There is also a privileged virtual environment, called dom0, which has access to physical devices, and provides access to the I/O operations from other domains. Management operations are also carried out by hypervisor. Figure 8.15 shows an example of Xen-based network virtualization. In this scenario, a virtual router migration is equivalent to migrating a virtual machine from a physical machine to another physical machine. As routers perform real-time service, a virtual router migration demands the minimization of the packet forwarding service downtime.

Xen native migration [CLA 05] is based on virtual machine live migration. The virtual machine live migration consists of copying the virtual machine memory from the source physical machine to the destination physical machine. This migration is called live because it is accomplished with the virtual machine running during the first phases of the migration and it reduces the virtual machine downtime. As the pages of virtual machine memory in the source physical machine may change during the migration procedure, this live procedure uses an iterative copy mechanism of memory pages. In the iterative copy, the modified memory pages are tagged, and, in the next iteration, they are copied to the destination. This is repeated until the number of modified memory pages in the last round is lower than a certain threshold. When the virtual machine execution is suspended in the source physical machine, the last modified memory pages are copied to the destination, and the virtual machine is, then, restored in the destination physical machine. A disadvantage of this proposal for virtual router migration is the packet loss during the time that the virtual machine is unavailable, i.e. during the elapsed interval between suspension and restoring. This mechanism, however, is limited to a local network, because this approach assumes the existence of a shared hard disk and link migrations are performed by sending *ARP Reply* packets.

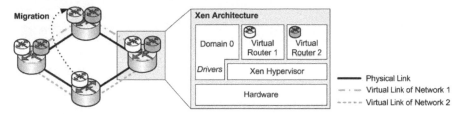

Figure 8.15. *Xen network virtualization. A network element migration is equivalent to a virtual router migration*

Plane separation can be used to avoid packet loss during migration. Pisa *et al.* propose a virtual machine migration for Xen platforms that makes a copy of the data planes of all virtual routers into dom0 [PIS 10]. Thus, the data plane migrations are performed without affecting routing and without downtimes. The solution, however, is restrictive since a virtual router migration is performed only between nodes in the same LAN. Hence, the migration scope is limited to just one hop from the source router.

8.2.2. *XenFlow architecture design*

The proposed architecture design combines the advantages of per-hop packet processing and distributed control, provided by Xen platform, with fast data flow processing capability, provided by OpenFlow platform. The architecture of a XenFlow network element is shown in Figure 8.16. Each virtual machine hosts the control plane of a different network. The OpenFlow switch in dom0 performs packet forwarding, according to the forwarding rules specified by the virtual machines. In this architecture, a network element can be defined as a virtual switch (layer 2), as a virtual router (layer 3) or as a middle box (layer greater than 3). The function performed by each network element depends on the virtual network protocol stack and applications.

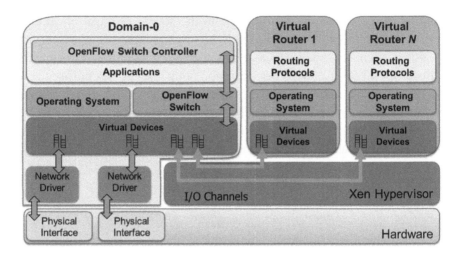

Figure 8.16. *Architecture of a XenFlow network element. An OpenFlow switch and an OpenFlow Controller run on Domain 0 of physical router*

In the XenFlow system, as well as in the Xen platform, physical device drivers are in dom0, and, then, all communications between virtual machines and physical devices traverse the dom0. Thus, the dom0 multiplexes packets from virtual network elements to physical devices and demultiplexes packets from physical devices to virtual network elements [FER 11]. In XenFlow architecture, the multiplexing and demultiplexing process is performed by an OpenFlow switch. Packet forwarding is programmed according to the rules defined by the OpenFlow switch controller, which is an application running in dom0. This controller interacts with the virtual machines to discover the forwarding rules created by each data plane. If a virtual machine (virtual network element) demands a per-packet processing, the controller will forward all incoming traffic of this virtual network to the virtual machine that runs the corresponding virtual network element. In addition to programming flows, this controller is also able to set policies for the flows, for instance, specifying a minimum bandwidth for each flow or for a set of flows [MCK 08].

In XenFlow architecture, a virtual network can select to work with a centralized or a distributed control plane. To establish a virtual network with distributed control, a virtual router is instantiated in each physical router (node) that belongs to that virtual network. Each virtual router interacts with the corresponding XenFlow controller in dom0 to inform the forwarding rules. However, in the centralized control mode, the XenFlow behaves like a OpenFlow switch. Hence, to establish a virtual network with centralized control, a middle box with a main controller is instantiated. This central controller interacts with individual XenFlow controllers that reside in the dom0 physical routers, which host the virtual network element. In this case, there is no need to instantiate a virtual machine in each physical node, in a model similar to traditional OpenFlow networks. Figure 8.17 shows the XenFlow architecture with three virtual networks, in which there are virtual switches and virtual routers interoperating. The physical switch data plane is shared using the OpenFlow protocol among different virtual networks.

Figure 8.17 shows the management entities that are aware of the physical network topology and that are managed by the infrastructure administrator, who decides when to instantiate or delete a virtual network, as well as deciding on the amount of physical resources that each virtual network consumes. These entities are also responsible for starting network migrations.

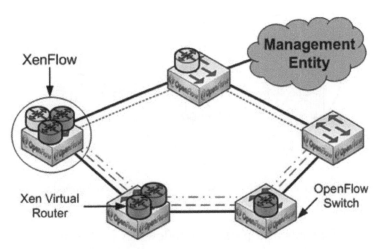

Figure 8.17. *Example of a XenFlow network composed of virtual switches and virtual routers*

8.2.2.1. *Plane separation and route translation into flows*

Flow migration is trivial when using an OpenFlow switch, but path migration is more complex on the Xen platform since turning off a router results in packet losses. Nevertheless, router migration without packet losses is fundamental and this goal is reached with XenFlow architecture because of the plane separation technique.

The forwarding table of each virtual network is computed by the virtual router, which takes place in a virtual machine. This information must be transmitted to dom0 so that the correct forwarding rules in the OpenFlow switch can be built. Figure 8.18 shows how XenFlow performs this task. A daemon running in the virtual machine copies the data plane information onto a Nox controller in dom0. Then, the controller translates these data into flows using the Rule Table module, which was developed as a Nox application, and configures the OpenFlow switch on demand.

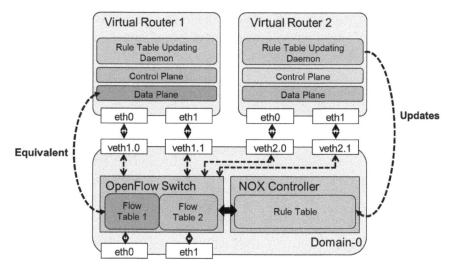

Figure 8.18. *XenFlow routing, in which packets are directly forwarded by Domain 0*

XenFlow forwards data packets as follows. A packet that reaches dom0 is directly forwarded, if it matches any flow in the Flow Table; otherwise, the packet is forwarded to the controller, in order to have its path defined by the controller. In this case, Nox controller extracts the 12 OpenFlow fields from the packet, queries the Rule Table to define the network that the packet belongs to and the corresponding forwarding rule of that network. After that, the controller inserts a new flow in the Flow Table of the OpenFlow switch. It is important to note that the packet arrives at dom0 with the destination Medium Access Control (MAC) address of the virtual router and this address has to be modified to the next hop MAC address before being forwarded by the OpenFlow switch. The next hop MAC address is obtained from the Rule Table. In case the node is a virtual switch, this operation is not performed. Hence, this module guarantees the correct mapping of a virtual link onto one or more physical links.

8.2.2.2. *XenFlow virtual topology migration*

In a XenFlow network, a virtual link can be mapped onto one or more physical links. The packet forwarding is accomplished by a flow table dynamically programmed by the Nox controller. Thus, the logical and the physical topology are detached. Therefore, virtual node migration in a XenFlow network, as shown in Figure 8.19, is composed of three steps: control plane migration, data plane migration and link migration. The control plane migration occurs from the origin to the destination physical node in a similar way to Xen conventional live migration mechanism [CLA 05]. After the control plane migration, all the data flows of the migrating virtual network must be mapped onto the new virtual topology. The old data path must be converted into a new data path in the destination physical router. Hence, the data plane migration is accomplished as follows: the data flow table entries related to the migrating virtual router are selected and transferred to the destination physical router; at the destination, these data flow definitions are mapped onto the current setup of the physical and virtual routers. Thus, the correspondence of flow input and output ports is kept, taking into consideration the dom0 virtual switches at the source and at the destination of the migration. Then, the corresponding flows of the migrated network are added to the Flow Table of destination dom0 OpenFlow switch. Finally, after data and control planes migration, the dom0 OpenFlow switches and the others network switches proceed the link migration. The link migration creates a switched path between the one logical hop virtual router neighbors and the destination physical router. Flows are thus defined in the physical routers that are in the path between the destination physical router and the physical routers, which host the virtual routers that are one hop away from the migrated virtual router. An automatic mechanism is required to create, on demand, new flows on physical routers along the path. This mechanism is used by the introduction of new rules on controller Rule Tables of the nodes in the path.

8.2.3. *Experimental results*

A prototype of the proposed architecture was developed as a proof of the concept of virtual routers' migration without packet loss. To evaluate the performance, we used the tool *Iperf* [IPE 13], as a packet generator tool, and *Tcpdump* [TCP 13] to measure the amount of packets generated, received and lost. The packet loss was evaluated from the information collected using *Tcpdump* on network interfaces responsible for generation and for reception of packets. To assess XenFlow performance, migration in XenFlow is compared to the native migration of the Xen virtualization platform.

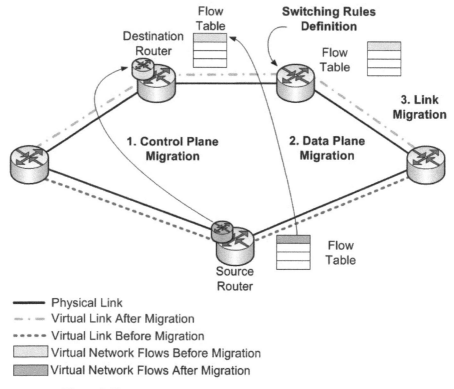

Figure 8.19. *Three steps of the XenFlow virtual topology migration*

The experimental scenario is composed of four machines: two for packet forwarding and two for generate/receive packets. Two machines perform the function of forwarding packets, and the XenFlow prototype was installed in these two machines. Each machine is equipped with an Intel Core 2 Quad processor and three Ethernet network interfaces of 1 Gb/s. The machines run the Xen hypervisor 4.0-amd64. In one of these physical machines, a virtual machine is instantiated with one virtual CPU, 128 MB of memory, two network interfaces and Debian 2.6-32-5 operating system to work as a virtual router. The experiments use two other machines, equipped with Intel Core 2 Duo processors, that generate or receive packets, each one equipped with an Ethernet interface 1 Gb/s, connected to a control network, and two Ethernet network interfaces 100 Mb/s, to communicate with both physical routers. In the experiments, virtual routers forwarded UDP packets of 64 and 1,500 bytes, which are, respectively, the minimum and maximum data length (maximum transmission unit (MTU)) of an Ethernet frame, respectively.

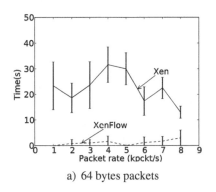

a) 64 bytes packets

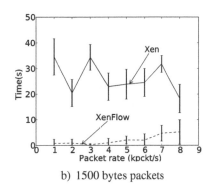

b) 1500 bytes packets

Figure 8.20. *Control plane downtime*

The first experiment measures the control plane downtime during migration. We send control packets that are forwarded by the virtual machine to determine the average downtime according to the number of lost control packets. Hence, the control plane downtime is given by the difference between the time stamp of the last packet received immediately before the migration, and the time stamp of the control packet received immediately after migration. Figure 8.20 shows the control plane downtime for the XenFlow system and for the native Xen migration, as a function of the transmitted packet rate. Results show that the average downtime of the virtual router is always lower than 5 s in XenFlow, regardless of the packet size. In native Xen migration, however, the average downtime of virtual router is always greater, ranging from 12 to 35 s. This difference mainly occurs for two reasons. First, there is no writing on the virtual machine memory during the migration using XenFlow, because the packets are sent directly by dom0, while in Xen migration, all packets are forwarded by the virtual machine. Hence, packet forwarding in Xen generates memory writings and readings while the virtual machine is migrated. These writing and reading memory operations lead to a greater number of dirty pages and, consequently, large downtime when copying the last memory pages of the virtual machine. Moreover, XenFlow migration is performed in two steps. First, the virtual machine is migrated and then the data plane is migrated using OpenFlow, avoiding packet losses. On native Xen migration, the link migration is accomplished by sending *ARP Reply* packets, in order to indicate the interfaces in which the migrated virtual machine is available. The *ARP Reply* mechanism is, however, based on the expiration of the ARP tables entry, which can add delays in updating the interfaces used to communicate with the migrated virtual machine.

The second experiment evaluates the total migration time, which establishes the minimum time between two consecutive virtual network element migrations. The total migration time considers the execution time of all operations related to the migration process. Figure 8.21 presents the total migration time as a function of the transmitted packet rate. Results show that XenFlow migration demands greater total migration time than Xen does. This result is expected because XenFlow migration involves more steps, and one of them is the native Xen migration itself. Part of the additional time occurs due to flow migration, given the need to rebuild the data plane at the destination physical router, and also due to link migration, which is responsible for setting up the new topology of the virtual network over the physical network. Figure 8.21b) shows that for 1,500 bytes packets, there is an increase in XenFlow total migration time, as the transmitted packet rate increases. This is caused due to the fact that, when using packets of 1,500 bytes, the link of 100 Mb/s is saturated under the rate of approximately 8,000 packets per second.

The third experiment shows the number of packet losses during the migration in both systems. Figure 8.22 reveals that, while migrating a router using XenFlow, there is no packet loss. Packet loss generated by Xen native migration increases with transmitted packet rates because the losses are due to the forwarding service downtime, as shown in Figure 8.20. As the native Xen migration downtime is almost constant, the amount of lost packets in this interval increases with the sending rate.

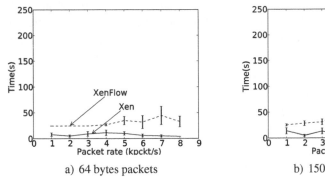

a) 64 bytes packets

b) 1500 bytes packets

Figure 8.21. *Total migration time*

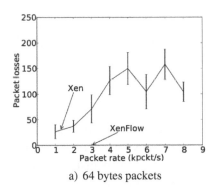
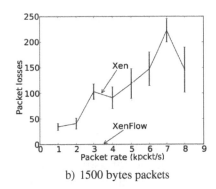

a) 64 bytes packets

b) 1500 bytes packets

Figure 8.22. *Amount of packet losses as a function of the
transmitted packet rate*

8.3. Summary

In this chapter, an overall view of developed architectures is presented. The first architecture presented is based on the Xen virtualization platform. We developed a model that integrates some developed features, in addition to new control and management modules. The main modules in the Xen architecture are the resource managers, which deal with the allocation of physical resources to virtual networks according to the policies previously defined. A fuzzy controller is also presented for observing the use of shared bandwidth, CPU and memory in dom0. The controller handles the resource consumption in the virtual machines, and interacts with the Xen scheduler to ensure the bandwidth provision for virtual machines. To integrate these modules, monitoring functions were modified by the proposed module and ADAGA. Other modules were also added, such as the admission controller, which evaluates whether new virtual networks can be hosted in a specific physical node. The admission controller is the basis for applying global functions that remap virtual networks according to network load. A QoS module for the proposed architecture was also introduced to accomplish the requirements of a new architecture for the Internet.

Tests were carried out to evaluate the performance of the new modules added to the Xen architecture. Results show that the developed mechanisms perform better than existing mechanisms in the literature. In addition, results also show that the proposed admission control dynamically adapts itself to different traffic demands, guaranteeing an accurate control of the number of virtual networks hosted in the physical device. Both resource overload and under-utilization were avoided, which differs from other proposals analyzed. Tests were also conducted with the developed QoS module. Results show the advantages of providing QoS among virtual networks in comparison to guaranteeing QoS primitives only inside the virtual network.

The second developed architecture uses the OpenFlow technology and the virtualization tool, called OMNI. Since OpenFlow network control is centralized, the infrastructure control modules were placed inside a special controller that interacts with FlowVisor for controlling physical resources provided to each virtual network. Modules developed for the Xen platform are added to control the resources in OpenFlow networks. Another functionality added to OMNI is the agent control, which emulates distributed control. The performed analysis shows the advantages of using agents as well as the impact of our management system on the virtualized network.

We also designed a third approach that capitalizes on the advantages of Xen and OpenFlow in a new virtualization platform. This approach is a hybrid network architecture design, which combines the flexibility of the OpenFlow switching matrix with the distributed control of a network virtualization based on Xen platform. This approach is called XenFlow and implements the concept of flow processing, in which a general purpose network node is able to process any kind of flow. XenFlow provides a robust and efficient way for migrating virtual topologies. The main goal of XenFlow is to achieve zero packet loss in virtual router migration and to eliminate the need for tunnels or external mechanisms for migration of links. The architecture design uses the plane separation technique used as an application over the Nox controller, which controls the data plane based on forwarding rules. These rules are updated by a daemon that runs in each virtual router. Results show that the XenFlow control plane downtime is up to 30 times lower than the native Xen migration downtime. Results also show that the total migration time of XenFlow is greater than the native Xen migration. These results occur due to new steps introduced by XenFlow into the virtual topology migration process, when compared with that of Xen. However, increasing the total migration time is not a significant factor for the virtual router migration. It only sets the minimum time between two consecutive migrations. Results show that a XenFlow virtual router migration occurs without packet loss, which makes this architecture design appropriate to the scenario of virtual networks, as opposed to the native migration of the Xen virtualization platform.

These network virtualization tools evolved to Future Internet Testbed with Security [FIT 13], which is used to experiment with new future Internet proposals [GUI 13].

8.4. Bibliography

[ALK 11] ALKMIM G., BATISTA D.M., DA FONSECA N.L.S., "Optimal mapping of virtual networks", *Proceedings of the IEEE Global Telecommunications Conference (IEEE Globecom 2011)*, IEEE, Houston, TX, pp. 1–6, 2011.

[ALK 12] ALKMIM G., BATISTA D.M., DA FONSECA N.L.S., "Approximated algorithms for mapping virtual networks on network substrates", *Proceedings of the IEEE International Conference on Communications (ICC 2012)*, IEEE, Ottawa, Canada, pp. 1–55, June 2012.

[BOL 09] BOLLA R., BRUSCHI R., DAVOLI F., *et al.*, "Energy-aware performance optimization for next-generation green network equipment", *Proceedings of the 2nd ACM SIGCOMM Workshop on Programmable Routers for Extensible Services of Tomorrow*, ACM, pp. 49–54, 2009.

[CAR 12] CARVALHO H.E.T., FERNANDES N.C., DUARTE O.C.M.B., "SLAPv: a service level agreement enforcer for virtual networks", *Proceedings of the International Conference on Computing, Networking and Communications, Internet Services and Applications Symposium (ICNC 2012)*, Maui, Hawaii, USA, pp. 1–5, January 2012.

[CLA 05] CLARK C., FRASER K., HAND S., *et al.*, "Live migration of virtual machines", *Proceedings of the 2nd Conference on Symposium on Networked Systems Design & Implementation-Volume 2*, USENIX Association, Anaheim, CA, EUA, pp. 273–286, April 2005.

[COU 11] COUTO R.S., CAMPISTA M.E.M., COSTA L.H.M.K., "XTC: a throughput control mechanism for Xen-based virtualized software routers", *Proceedings of the IEEE Global Communications Conference (GLOBECOM 2011)*, Houston, TX, USA, pp. 2496–2501, December 2011.

[EGI 07] EGI N., GREENHALGH A., HANDLEY M., *et al.*, "Evaluating Xen for router virtualization", *International Conference on Computer Communications and Networks (ICCCN07)*, Honolulu, Hawaii, USA, pp. 1256–1261, August 2007.

[FAJ 11] FAJJARI I., AITSAADI N., PUJOLLE G., *et al.*, "VNE-AC: virtual network embedding algorithm based on ant colony metaheuristic", *Proceedings of the ICC 2011 Next Generation Networking and Internet Symposium (ICC11 NGNI)*, IEEE, Kyoto, Japan, pp. 1–6, June 2011.

[FEA 07] FEAMSTER N., GAO L., REXFORD J., "How to lease the internet in your spare time", *ACM SIGCOMM Computer Communication Review*, vol. 37, no. 1, pp. 61–64, 2007.

[FER 11] FERNANDES N., MOREIRA M., MORAES I., *et al.*, "Virtual networks: isolation, performance, and trends", *Annals of Telecommunications*, vol. 66, no. 5–6, pp. 339–355, Springer, Paris, 2011.

[FER 11a] FERNANDES N.C., MOREIRA M.D.D., DUARTE O.C.M.B., "XNet-Mon: a network monitor for securing virtual networks", *Proceedings of the IEEE International Conference on Communications (ICC 2011-Next Generation Networking and Internet Symposium)*, ICC'11 NGNI, Kyoto, Japan, 2011.

[FER 12] FERNANDES N.C., DUARTE O.C.M.B. VIPER: fine control of resource sharing in virtual networks, GTA Technical Report GTA-12-02, Electrical Engineering Program, COPPE/UFRJ, January 2012.

[FIT 13] Future Internet Testbed with Security project, available at http://www.gta.ufrj.br/fits/ (accessed in May 2013).

[GUD 08] GUDE N., KOPONEN T., PETTIT J., *et al.*, "Nox: towards an operating system for networks", *ACM SIGCOMM Computer Communication Review*, vol. 38, no. 3, pp. 105–110, 2008.

[GUI 13] GUIMARÃES P.H.V., FERRAZ L.H.G., TORRES J.V., *et al.*, "Experimenting content-centric networks in the future Internet testbed environment", in *Workshop of Cloud Convergence: Challenges for Future Infrastructures and Services (WCC-02)* – ICC'2013, Budapest, Hungary, June 2013.

[IPE 13] IPERF, 2013. Available at http://iperf.sourceforge.net/ (accessed in May 2013).

[MAT 11] MATTOS D.M.F., FERNANDES N.C., DA COSTA V.T., *et al.*, (accessed in May 2013). "OMNI: OpenFlow management infrastructure", *Proceedings of the 2nd IFIP International Conference Network of the Future – NoF2011*, Paris, France, pp. 53–57, November 2011.

[MCK 08] MCKEOWN N., ANDERSON T., BALAKRISHNAN H., *et al.*, "Openflow: enabling innovation in campus networks", *ACM SIGCOMM Computer Communication Review*, vol. 38, no. 2, pp. 69–74, 2008.

[PIS 10] PISA P., FERNANDES N., CARVALHO H., *et al.*, "Openflow and Xen-based virtual network migration", in PONT A., PUJOLLE G., RAGHAVAN S. (eds), *Communications: Wireless in Developing Countries and Networks of the Future, vol. 327 of IFIP Advances in Information and Communication Technology*, Springer, Boston, pp. 170 181, 2010.

[PIS 11] PISA P.S., COUTO R.S., CARVALHO H.E.T., *et al.*, "VNEXT: virtual network management for Xen-based testbeds", *Proceedings of the 2nd IFIP International Conference Network of the Future – NoF2011*, pp. 41–45, Paris, France, November 2011.

[RAT 05] RATNASAMY S., SHENKER S., MCCANNE S., "Towards an evolvable internet architecture", *ACM SIGCOMM Computer Communication Review*, vol. 35, no. 4, pp. 313–324, 2005.

[ROD 12] RODRIGUEZ E., ALKMIM G., BATISTA D.M., *et al.*, "Green virtualized networks", *Proceedings of the IEEE International Conference on Communications (ICC 2012)*, IEEE, Ottawa, Canada, pp. 1–6, June 2012.

[SHE 09] SHERWOOD R., GIBB G., YAP K., *et al.*, Flowvisor: a network virtualization layer, Technical report, Technical Report OPENFLOW-TR-2009-01, OpenFlow Consortium, 2009.

[TCP 13] TCPDUMP & LIBPCAP, 2013. Available at http://www.tcpdump.org/ (accessed in May 2013).

[WAN 07] WANG Y., VAN DER MERWE J., REXFORD J., "VROOM: virtual routers on the move", *Proceedings of the ACM SIGCOMM Workshop on Hot Topics in Networking*, Atlanta, Georgia, USA, November 2007.

[WAN 08] WANG Y., KELLER E., BISKEBORN B., *et al.*, "Virtual routers on the move: live router migration as a networkmanagement primitive", *Proceedings of the ACM SIGCOMM*, pp. 231–242, August 2008.

[WOO 09] WOOD T., SHENOY P., VENKATARAMANI A., *et al.*, "Sandpiper: black-box and gray-box resource management for virtual machines", *Computer Networks*, vol. 53, no. 17, pp. 2923–2938, 2009.

List of Authors

Miguel Elias M. CAMPISTA
Universidade Federal do Rio de Janeiro
Brazil

Luís Henrique M.K. COSTA
Universidade Federal do Rio de Janeiro
Brazil

Otto Carlos M.B. DUARTE
Universidade Federal do Rio de Janeiro
Brazil

Nelson Luís S. DA FONSECA
Universidade Estadual de Campinas
Brazil

Edmundo R.M. MADEIRA
Universidade Estadual de Campinas
Brazil

Igor M. MORAES
Universidade Federal Fluminense
Niterói
Brazil

Guy PUJOLLE
LIP6
UPMC
Paris
France

Index